Music in Orbit

Music in Orbit

SATELLITE RADIO IN
THE STREAMING SPACE AGE

Brian Fauteux

UNIVERSITY OF CALIFORNIA PRESS

University of California Press
Oakland, California

© 2025 by Brian Fauteux

All rights reserved.

Cataloging-in-Publication Data is on file at the Library of Congress.

ISBN 978-0-520-41415-0 (cloth)
ISBN 978-0-520-41416-7 (pbk.)
ISBN 978-0-520-41417-4 (ebook)

34 33 32 31 30 29 28 27 26 25
10 9 8 7 6 5 4 3 2 1

For Dallas, Felix, and Beatrix

Contents

	Acknowledgments	ix
	Introduction	1
1.	The Songs Down to Earth: Music Radio and Outer Space	28
2.	Targeting Subscribers in Satellite Radio's Formative Years	60
3.	Perceived Value and Satellite Streams in Music Programming	95
4.	Disposable Music and Monopoly Power in Policymaking	129
5.	The Stars Down to Earth: The Celebrity Radio Voice on Satellite Radio	155
6.	The Transnational, Technological, and Programming Expansion of SiriusXM	184
7.	Embedded Radio and the Limits of Expansion	208
	Notes	233
	Bibliography	279
	Index	293

Acknowledgments

The other day I was driving to the Rocky Mountain town of Canmore for the weekend. Heading west, the setting sun was turning the sky a wonderful mix of blue, purple, and orange. On the radio, Cindy Lee's "If You Hear My Crying" was playing. It's a song off one of the most critically acclaimed albums of 2024 so far, *Diamond Jubilee*, and one that has sparked generous conversation about the nature of music in the streaming era, given that it is (refreshingly, in my opinion) unavailable on streaming services and instead is available for purchase via an unappealing Geocities website and an e-transfer of money. It ties in well with the theme of this book, just as it tied in well with the serene mountain drive.

I use this example because I was listening to *Gorilla vs. Bear Blog Radio* on SiriusXMU, a satellite radio channel programmed by satellite radio company SiriusXM. Satellite radio is ideal for long drives because you don't lose reception in the same way that you do with broadcast radio. I listened to SiriusXMU ten years ago while driving from Montreal, Quebec, to Madison, Wisconsin, to begin a postdoc where the research for this book began. Across these locations and over about a decade or so, satellite radio has been a consistent presence during a period that was often anything but. I enjoy the music on SiriusXM, and at the same time I've watched (and listened) as concentration and consolidation in the satellite radio industry, and music industries more broadly, have resulted in more limited playlists and less experimental programming (though gems like *Blog Radio* remain). My writing on this topic reflects the meaningful ways that music and music radio are embedded in my daily life.

My daily life has been eventful while writing this book, namely due to my wonderful children, Felix and Beatrix, whose boundless energy and happiness have inspired me to sit, write, and focus when the time has been right. To Dallas, you've been as constant as the Northern Star, the brightest light that shines, (to borrow a line from Gerry Rafferty), and to my wider constellation of dear family members in the Fauteuxs and Curows, I can't thank you enough for the constant love and support.

This book has benefited greatly from conversations with friends, colleagues, mentors, and students. My great friend and regular research collaborator, Andrew deWaard, has helped me think through my ideas behind this book for a long time and has offered careful and thoughtful comments on my writing. Brianne Selman, who along with Andrew comprises The Cultural Capital research team, has also been a source of insight into the world of music industries research on distinct but related work. I'm truly lucky to work among excellent and collegial colleagues in the Department of Music at the University of Alberta. Thank you to my chair, Bill Street, for the support of this work since I arrived here in 2015, and especially to Julia Byl, Patrick Nickleson, and Fabio Morabito, who have each either read and offered thoughtful comments on sections of this book or at least have listened to me talk about it over wings and beer. I am very thankful for the excellent research assistance I received over the years from a number of University of Alberta students: Matthew Green, Chloe Carpenter, Aanchel Gupta, Piper Keyes, Breezy Prochnau, and Sydney Horrocks.

My interest in satellite radio and music began while I wrapped up my PhD at Concordia University in Montreal. At that time, I was encouraged by my exceptional supervisors and mentors, Charles Acland and Leslie Shade, to apply for a postdoc and to approach the University of Wisconsin–Madison as a host institution. Thanks especially to Charles for ongoing conversations about this project and about music and for helping me think through where to publish this book. Thank you to Jennifer Holt, as well, for weighing in on this discussion. At the University of Wisconsin, I was fortunate to have Michele Hilmes as a host supervisor for the early stages of this project and to spend two years with a wonderful group of faculty members and graduate students. Thanks to all of you, and particularly to Jeremy Morris, Jonathan Gray, Evan Elkins, Kit Hughes, Josh Shepperd, Andrew Bottomley, Christopher Cwynar, Alyx Vesey, Myles McNutt, Nora Patterson, and Jennifer Wang for being friends and colleagues both during my time in Madison and in the years since. Many of you also provided feedback on earlier stages of this satellite radio research.

In annual conferences hosted by associations including SCMS, IASPM (Canada and US), and the CCA, I have found a great network of music and radio scholars who have contributed much to my thinking on music and radio and who have been in the same conference room as me more times than I can count. Some of you are already listed above; others are Alexander Russo, Elena Razlogova, Andy Stuhl, Amy Skjerseth, Jason Loviglio, Eric Harvey, Gregory Taylor, Bill Kirkpatrick, David Madden, Catherine Martin, and Ilana Emmett. At one of my first conference presentations on this research in Victoria, B.C., I was the only panel member to show up, and there were three audience members in the room, including Richard Sutherland and Line Grenier. Without a doubt, the comments I received from both of you after that presentation have been some of the most helpful I've ever gotten on a conference presentation (quality over quantity, I always have to say now). A huge thanks to Alex, Elena, Jason, and Lisa Parks for the generous blurbs that now accompany this book.

Thank you to Raina Polivka, Sam Warren, and the entire UC Press team for the guidance and support. Thank you to the reviewers of my draft manuscript for the focused and supportive comments; the final version is in much better shape thanks to you both. I'm grateful to both Jennifer Waits and Jason Loviglio for happenchance conversations that pointed me to resources and interview subjects I hadn't thought of. My gratitude to Jennie Thomas at the Rock & Roll Hall of Fame Library and Archives. Thanks to the interviewees who spoke with me as well as those who initially wanted to but ultimately couldn't.

Certain parts of chapter 5 have been published in *Popular Music* ("When the Dial Goes Dylan: 'Premium' Radio, Hybrid Authenticity, and *Theme Time Radio Hour*") and in the *Journal of Radio and Audio Media*

("Blog Radio: Satellite Radio and the Aesthetics of Podcasting"), and aspects of chapter 6 have been published in the *International Journal of Cultural Policy* ("Satellite Footprint to Cultural Lifelines: Sirius XM and the Circulation of Canadian Content"). My gratitude to these journals, their editors, and reviewers who helped shape my thinking and writing as I worked on these articles.

And thank you to the Social Sciences and Humanities Research Council of Canada for the postdoctoral research funding that began this journey.

Introduction

The number *one*. In music, it's a marker of success, popularity, and chart dominance. Propelled by radio and its cadre of influential hosts and DJs, a number one hit solidifies an artist's influence and relevance. Sometimes a number one reverberates across generations; at other times it lingers for a week or two, a fleeting moment in time still charged with cultural resonance. In the twenty-first century the influence of radio on hitmaking remains, but it shares this role with music videos, streaming music playlists, and user-generated posts on social media platforms. Number-one songs are increasingly global, orbital, cross media, and more immersive. How and why does a number-one song generate such value, and how do songs circulate and impart that value in our lives, soundtracking both the magical and the mundane? Further, how do media industries shape that circulatory negotiation of culture and community?

Two massively successful albums released at the turn of the twenty-first century demonstrate the importance of the number one. The first, The Beatles' *1*, includes a total of twenty-seven songs that hit number one in the UK or the US, and it was released on November 13, 2000. The other, Elvis Presley's *ELV1S: 30 #1 Hits*, was released by RCA Records on September 24, 2002. With thirty hit songs, *ELV1S* has a runtime of nearly eighty minutes and was released in time to celebrate the twenty-fifth anniversary of the rock star's death. These compilation albums, full of number-one hits, fortify the legacies of these artists and signal the undeniable staying power of hit songs in the popular music industries. Early hits on each album include The Beatles' "Love Me Do" (released in the UK in October 1962) and Elvis's "Heartbreak Hotel" (January 1956). The Beatles album includes songs that hit number one on the *Record Retailer* Top 50 chart or the *Billboard* Hot 100 chart.[1] The Elvis album draws from the pop singles charts of *Billboard*, *Cashbox*, *Record Retailer*, and *New Musical Express*.[2] Both compilation albums also hit the top of the *Billboard* 200 charts for a substantial amount of time (The Beatles for eight weeks and Elvis for three). The Beatles' *1* was the highest-selling album of the 2000s, and *ELV1S* was certified six times platinum in 2018. As music critic Tom Breihan writes, hit songs, like those of the Hot 100, "chronicle each new trend that would capture the collective imagination: the Beatles, Motown, psychedelia, disco, new wave, metal, rap."[3] These songs and albums are not just about commercial success but also about shaping the lives of listeners on personal and collective levels. Popular songs become a common reference point across time. They tell us much about the media, culture, and technology of the moment at which hit songs emerge or reemerge.

The Beatles and Elvis had something else in common at the dawn of the new millennium: their own dedicated satellite radio channels. First to launch was Elvis Radio, on Sirius Satellite Radio in 2004, right between the release of the number-one hits album and the merger of then-competing satellite radio companies Sirius and XM (the merger was finalized in 2008). The Beatles Channel debuted on May 18, 2017, on SiriusXM, the now-merged satellite radio service that has monopoly status in the uniquely small satellite radio industry. Both of these channels showcased the staying power of these legacy artists. Further, they reflected one of SiriusXM's key programming tenets: circulating music from the past, often in a way that goes *beyond* the most radio-friendly number-one hits, mining deeper into catalogs to fill a twenty-four-hours-a-day, seven-days-a-week, artist-based channel that is exclusive to satellite radio subscribers. In 2023 artists with their own SiriusXM channel included Pearl Jam, Phish, the Grateful Dead, and Bob Marley. A pop-up channel celebrating the music and life of Tina Turner was active for a limited time as

I wrote this sentence. In order to fill a full day of programming, listeners are hearing more than the top hits on these channels. Rather, each channel has its own idea of what a hit song might be for the specific listeners who are tuning in. It may be one closely aligned with chart success, or it might involve an array of songs that are deep cuts, or perhaps those that reflect the breadth and depth of an artist's career. A song by Slayer is a hit on the metal channel but "meaningless" on the Top 20 channel.[4]

The first attempt to establish satellite radio was made by a company named WorldSpace. Its founder, Noah Samara, had the idea for satellite radio in 1990, a service that used satellites with three beams, each delivering more than fifty channels to proprietary radios. WorldSpace covered seventy countries, with a potential listenership of 4.6 billion. Its coverage included Africa, the Middle East, a large part of Western Europe, Japan, China, Southeast Asia, India, and Iran.[5] WorldSpace no longer exists after filing for bankruptcy in 2008. Another service, Global Radio in Luxembourg, planned to start in 2005.[6] Both Sirius and XM would be less ambitious in their satellite coverage: first the United States and then Canada, Puerto Rico, and northern parts of Mexico. The Washington, D.C.–based XM Satellite Radio launched the first American digital satellite radio service on September 25, 2001. *Time* named the new service the "invention of the year" and *Fortune* the "product of the year." Sirius Satellite Radio experienced a number of delays in rolling out its service and had its debut postponed to early 2002.[7] It was limited to four states in February and would expand nationally in July. Satellite radio came on the market at the start of a new decade, a new century, and a new millennium. By satellite radio, I specifically mean satellite digital audio radio service (SDARS), which is distinct from radio by satellite in that it's a unique digital radio service provided by satellite on the S-band (spectrum between 2320 and 2345 MHz). SiriusXM is now the only SDARS service in the world.

Just before the launch of the new services at the turn of the millennium, music culture was experiencing dramatic shifts and changes due to the ability for everyday users to easily rip and share music files, namely MP3s, over the internet. The recording industry was up in arms about this development, and the Recording Industry Association of America (RIAA) notoriously issued fines of thousands of dollars to people who may have downloaded a handful of songs.[8] In the world of radio, industry personnel were curious about what the future would hold when music could be freely found and heard online. Beginning in June 1999, the internet file-sharing application Napster had, as *Billboard* put it, "created a culture of teens and young adults who use [it] to discover cool music," placing the usefulness and appeal of radio in jeopardy.[9] This feeling of instability was also reflected in a moment of techno-dystopian fears of wide-scale computer problems and shutdowns, known as the Y2K problem, scare, glitch, or simply Y2K.

Despite the uncertainty in the record business, major pop groups were finding success on an increasingly global scale in the early 2000s. The world's best-selling album in 1999, the Backstreet Boys' *Millennium*, hit number one in twenty-six countries. Other albums with global success in 2000 included NSYNC's *No Strings Attached* (certified platinum or multiple times platinum in Argentina, Australia, Canada, Denmark, New Zealand, and America, selling nearly ten million copies in 2000 alone), Madonna's *Music* (hitting number one in multiple countries across the globe), Britney Spears's *Oops! . . . I Did It Again*, and Eminem's *The Marshall Mathers LP*. An advertisement for the Backstreet Boys' follow-up album to *Millennium*, *Black & Blue* (an album also from 2000 that, for a moment, knocked The Beatles' *1* out of the number-one spot on the *Billboard* 200), referred to the new album as an upcoming "global assault," with a lead single, "Shape of My Heart" that was "another international smash."[10] The band embarked on a media junket that would cover five continents in five days.

"Global assault" became an unfortunate and untimely phrase for promoting global hit music nearly two years into the new millennium when the September 11, 2001, terrorist attacks brought the United States, and much of the world, to a halt. In the world of radio, *Billboard* declared that the attacks shaped radio as it "changed priorities in a world that has been altered forever." Some radio stations turned to "inspirational and patriotic themes," and listeners used music on the radio as an escape from reality.[11] With an economic recession and heightened consolidation in the radio industry, many commercial broadcasters were also making significant layoffs.[12] Songs released around this time, like The Strokes' "New York City Cops" and Wilco's "Jesus, Etc.," had their meanings altered and tailored to the themes of the attacks and the broader historical context surrounding

Figure 1. Ad for the Backstreet Boys' album *Black & Blue*. Source: *Billboard*, November 25, 2000.

them, despite both songs having been written and released before September 11. Songs are malleable, and the most notable marks on history's time line can bend them to their will.

This book is about music and culture at the turn of the new millennium and over the last thirty years. It captures a time when certain aspects of music culture became increasingly centralized and imbued with further significance, legacy, and value, as the catalogs of The Beatles and Elvis demonstrate, while other aspects became increasingly fragmented, niche, weird, and personal, like online music communities and the move to more personal and individual modes of listening. It is about a time when concentration in the radio and record industry, and later in Big Tech, would accelerate and alter the ways listeners hear music. And it's about a time when the methods by which we pay for music, or don't pay for music, changed dramatically, moving from paying for recorded music as a material copy, like a record on vinyl or a CD, to paying for subscription access to a streaming music platform. Over the course of this transition, a media company that is less frequently discussed is SiriusXM. The satellite radio company and its earlier distinct companies (Sirius and XM) anticipated many of the cultural, musical, and technological shifts that now define the 2020s. Both Sirius and XM began by demonstrating their ability to offer *more* than terrestrial broadcast radio, radio *beyond* radio, and now currently, the merged SiriusXM under Liberty Media control stands as one of a few massive music companies that operate in the nexus of internet-streaming, online audio, and radio. These massive music companies are implicit in the ways music circulates and in the ways music is ascribed value, legitimacy, and currency in the contemporary music environment.

Early advertising campaigns for XM and Sirius leaned into the potential benefits that radio by satellite could offer. One XM television commercial from 2002 boasted a "revolutionary new kind of radio that delivers more choice, better sound, and coast to coast coverage." After showing a satellite flashing across outer space, the commercial depicts violins, a jukebox, a grand piano, and vinyl records raining down from the sky.[13] A 2005 Sirius Satellite Radio commercial that aired in Canada showcased its music, sports, and news programming and included one man closing his car doors and saying, "It's flat out better than regular radio."[14] Finally, one promotional video put together by radio historian Art Vuolo Jr. on the early days of XM shows DJs Phlash Phelps and Cleveland Wheeler performing a test run of XM radio on June 19, 2001. As "The Battle Hymn of the Republic" plays, they alter and announce their own take on the US Constitution, saying, "We the people of XM, in Order to form a more perfect radio, establish Justice, insure domestic Tranquility, provide unabridged entertainment for the masses." The sounds of bombs crashing and fireworks accompany their proclamation.[15] Further evidence of the sheen that XM was eager to showcase included leaning into the use of "X" as an edgy alternative to AM and FM radio. XM referred to "the Power of X" to delineate "the coming new wave of radio."[16] In its annual report for the year 2000, XM boasted "Technology to the Power of X; Radios to the Power of X; Experience to the Power of X; Music to the Power of X; Entertainment to the Power of X; Marketing to the Power of X; Management to the Power of X; Partnership to the Power of X."[17] The following year, XM proclaimed that it had rolled out America's first digital satellite radio service, "made it the fastest-growing new consumer electronics audio product in 20 years," and transformed "radio which [had] not had a technological change since the introduction of FM almost 40 years ago."[18] The company also took aim at the space program, arguing that it had done very little for "your average Joe," aside from inventing Tang and Velcro, until the advent of commercial satellites.[19]

Although both services positioned their attributes as exceeding the capabilities of broadcast radio, some writers indicated that XM was more advanced in its uniqueness. American media writer Douglas Rushkoff suggested that XM's superiority came from its efforts to reinvent radio and the business model "from the inside out," whereas Sirius applied the "strategies of other businesses to the specific case of a new technology and market."[20] Sirius would look to augment its offerings through massive talent contracts, such as signing Howard Stern to a five-year, $500 million deal in 2004. XM co-founder Lee Abrams reveals the sort of distinctiveness that XM was thought to have in its early years when he denounces the use of the word *service* for XM radio, a word he feels to be unexciting and unable to create fans out of radio listeners.[21] Satellite radio has been called an *audio service*, but Abrams views it as having been much more, a "living breathing national radio music powerhouse."[22] Regardless, for lack of a better term, I use the word *service* throughout this book to refer to

Satellite Radio At A Glance

Sirius   *XM*

Audio Makers

Sirius	*XM*
Alpine, Audiovox, Clarion, Delphi Delco, Eclipse, Jensen, JVC, Kenwood, Panasonic, Visteon	Alpine, Audiovox, Delphi, Pioneer, Sony

Auto Manufacturers

Sirius	*XM*
Audi, BMW, Chrysler, Dodge, Ford, Freightliner, Infiniti, Jaguar, Jeep, Lincoln, Mazda, Mercedes-Benz, Nissan, Sterling Trucks, Volkswagen, Volvo	Acura, Audi, Buick, Cadillac, Chevrolet, GMC, Honda, Infiniti, Isuzu, Nissan, Oldsmobile, Pontiac, Saturn, Toyota, Scion, Volkswagen

Consumer Electronics Retailers

Sirius	*XM*
ABC Warehouse, Audio Express, Best Buy, Brandsmart, Car Toys, Circuit City, Cowboy Maloney, Crutchfield, Good Guys, HH Gregg, Mickey Shorr, Mobile One, Nebraska Furniture Mart, P.C. Richard & Son, Sears, Sound Advice, Tweeter, Ultimate Electronics, Wal-Mart (expected), Avionics, Boater's World	Audio Express, Best Buy, Car Toys, Circuit City, Crutchfield, Good Guys, PC Connection, Tweeter, Ultimate Electronics, Wal-Mart

Out-of-Car Products

Sirius	*XM*
Plug & Play Unit, June 2003 Dedicated Home Units, September 2003	SKYFi Audio Receiver with home adapter kit SKYFi Audio System portable "boom box" XM Personal Computer Receiver

Current Ad Campaigns

Sirius	*XM*
NASCAR Sirius Dodge No. 7, driven by Jimmy Spencer, and two Sirius-branded races, billboards, direct mail, print, radio, TV	Direct mail, magazine, newspaper, online, radio, TV; plus GM ad campaigns promoting XM

Monthly Subscription Fee

Sirius	*XM*
$12.95, $6.99 for an add'l subscription	$9.95, $6.99 per radio for up to four add'l radios

	Sirius	*XM*
Subscribers as of 3/31/03	68,059	500,000
Targeted Subscriptions by 12/31/03	300,000	1.2 million
Targeted Quarter to Break Even	Early 2005 (based on 2 mil. subscribers)	Late 2004 (Cash-flow break-even)
First-Quarter 2003 Revenue	$1.6 million	$13.05 million
Market Capitalization (as of 5/28/03)	$1.267 billion	$1.207 billion

Figure 2. "Satellite radio at a glance" comparison chart. *Source:* Marc Schiffman, "Pie in the Sky: After Near Crashes, Satellite Radio Set to Soar," *Billboard*, June 7, 2003, 82.

satellite radio as a form of entertainment that is accessible through subscription fees.

Claims about the superiority of satellite radio, bold as they were, made sense for these new services at the time. Across the landscape of North American popular music in the early 2000s, a notable range of music styles was finding airplay on commercial stations. In addition, a growing number of bands and artists were also finding success and critical acclaim outside the boundaries of commercial broadcast radio. The media landscape was ripe for new players with more bandwidth. Artists who topped the *Billboard* 200 list in 2001 included Jennifer Lopez, Shaggy, Dave Matthews Band, 2Pac, Janet Jackson, Destiny's Child, Tool, Blink-182, and Garth Brooks. And while commercial broadcast radio was certainly supportive of these artists, playing their hit singles frequently and indicating the promotion of multiple mainstreams on the radio,[23] some bands and genres were generating buzz on the internet. In the growing indie scene in Canadian cities like Toronto and Montreal at the turn of the millennium, which included artists such as Broken Social Scene, Feist, and the Constantines, bands felt unsupported by labels and radio in the nation and instead looked for support and promotion in the US or Europe.[24] This was part of a broader problem of commercial radio deregulation and radio industry concentration that didn't favor independent artists or labels.[25]

Sirius and XM were also promoted as being distinct from commercial broadcast radio in their support for genres like hip hop and electronic dance music (EDM), as well as for an expansive list of subgenres of rock and pop. In the bandwidth of a satellite radio signal, one unscrambled by radio receivers to generate over one hundred channels, not only could a wider variety of genres or formats be included on the channel lineup, but so too could sub- or micro-genres. The satellite, then, becomes a figure for communicating progress and expansion, driving an ongoing myth or assumption about new media, which is that they foster innovation, an assumption we would be wise to complicate and challenge. For instance, this emphasis on music variety and distinction from commercial broadcast radio was certainly not the invention of the satellite radio industry. Sirius and XM borrowed from, and repackaged, the sonically diverse sounds of independent freeform FM radio of the 1960s and 1970s and, subsequently, college radio of the 1980s and 1990s (and beyond). Only now it was targeted to affluent listener-subscribers in North America.[26] But the satellite as a future-oriented technology is a further point of distinction against the idea of radio growing old, stale, and unable to grapple with the fact that listeners were increasingly able to locate and listen to a wide variety of music in the early 2000s. In many of the critically acclaimed albums and songs from the early millennium, we can hear traces of outer space, or of thinking beyond the terrestrial, in music. I expand on a few of these later; others include Spiritualized's *Ladies and Gentlemen We Are Floating in Space* (1997), No Doubt's *Return to Saturn* (2000), Air's *Moon Safari* (1998), Modest Mouse's *The Moon & Antarctica* (2000), the Beastie Boys' "Intergalactic" (1998), Monster Magnet's "Space Lord" (1998), Missy Elliott's "Gettaway" (1997), and Neko Case's "I Wish I Was the Moon" (2002).

On the album *In the Aeroplane Over the Sea* (1998), Neutral Milk Hotel strikes a unique balance between lo-fi simplicity and an expanded sonic atmosphere that evokes experimentation and draws on a wide range of instruments to signal the sort of instability and uneasiness that characterized the turn of the millennium. At the same time, it denotes the rising profile of indie rock and music culture outside of mainstream gatekeepers like commercial radio. Writing for *Pitchfork* in 2005, Mark Richardson explained that the album has a "disorienting relationship to time." This is because the album's "instrumentation seems plucked randomly from different years in the 20th century: singing saws, Salvation Army horn arrangements, banjo, accordion, pipes."[27] On the album's twenty-fifth anniversary, music critic Carl Wilson called it a "millennial identity marker," referencing its lasting legacy over two and a half decades.[28] Wilson referred to a *Vulture* article supposedly written by *Parks and Recreation*'s monotone millennial archetype April Ludgate, who described the album as a "bunch of weird, like, musical inspirations that flew down from outer space."[29] Musically and lyrically these elements come to the fore on the album's title track, on which the pairing of the acoustic guitar and musical saw delineate an otherworldly sound, one evoking a haunting sense of outer space, especially given the lyrics, "What a beautiful face / I have found this place / That is circling all 'round the sun." Sounds of the future and of space come through on a number of other tracks and albums at the turn of the millennium as well. Timbaland's production work on Aaliyah's hit 2000 song "Try Again" has been described as skewing "toward the fuzzy electronic sounds of the future,"[30] and the song wraps up with the lyric, "The new millennium, baby."

Although we can hear themes of new directions or the future in certain releases at the turn of the millennium, this is also a moment defined by a return to the past, or more precisely, the recycling of sounds and styles that had come before. Historian Lynn Hunt has said that time feels like a defining feature of our lives, and technologies like satellites and mobile phones "have made the experience of simultaneity an even more widely shared one."[31] Simon Reynolds has derided a cultural turn in which the new millennium's first decade has brought forth "every other previous decade happening again all at once: a simultaneity of pop time that abolishes history while nibbling away at the present's own sense of itself as an era with a distinct identity and feel."[32] Of course, actual culture, and music, is more complicated than this, and Reynolds, writing at the close of this first decade, was without the retrospective long-running glance backward that often elucidates what indeed was novel and curious about an era or a generation.[33] That said, the amalgamation of decades and genres and styles of music in this new radio service at the turn of the millennium certainly falls within his critique. The past becomes accessible through myriad niche channels. Even the mootness of the Y2K panic, whereby the ultimate collapse of global computer systems decidedly did not happen, indicates a theme of failure for the turn of the millennium to generate its own novel stories. The September 11 attacks on the World Trade Center, however, would bring forth a unique global moment that would set a distinct course in geopolitics, culture, and the economy, a trajectory within which satellites play a significant role both in communications and in military surveillance.

Satellite radio must be understood as a product of its time. The world of music was in the midst of changes in how music was listened to and distributed, and it's only fitting that this also altered the radio industry. If listeners were becoming accustomed to more choice, with access to seemingly endless music selections on the internet, it only makes sense that new developments in the radio industry would mirror these values. As the senior VP of content for Sirius said in 2000, radio has "pretty much calcified [over] the last 30 years. It's been over-consulted and over-consolidated.'"[34] This feeling of stasis toward the radio industry helped to justify the $1.5 billion that each company would pay to launch its new satellite radio service.[35] Claims of overconsolidation followed in the wake of the 1996 Telecommunications Act, which increased the number of stations one company could own and dampened the Federal Communications Commission's (FCC's) aims of promoting competition, diversity, and localism, which are stated in the American Communications Act of 1934. The marketplace theory and deregulation promoted by the 1996 act meant that a single entity could now own up to eight stations in a market and an unlimited amount nationwide. Following the implementation of the act, between March 1996 and March 2002 the number of radio stations increased, but ownership decreased 33.6 percent, and the two largest radio chains grew from fewer than sixty-five stations to more than fourteen hundred; Clear Channel's ownership increased 3,288 percent in seven years.[36] Consolidation in the radio industry has made it difficult for artists to get radio airplay, particularly those without the backing of big labels and the personnel who can lobby for radio play and promotion.[37] One *Future of Music* report on radio deregulation from 2002 explained that Clear Channel and Viacom controlled 42 percent of listeners and 45 percent of industry revenues at that time. With fewer companies controlling what music is played on the radio, there was a decrease in the diversity of music heard on air. Format *variety* had increased, but this simply points to the average number of formats available in each market. Deceptively, many distinct formats shared playlists or portions of playlists, with a playlist overlap as high as 76 percent in some cases. This problem was further exacerbated by concentration in the record industry, with only five major labels responsible for 80 to 100 percent of songs on radio charts (in 2023, there were merely three big labels: Sony, Universal, and Warner).[38] But as this book shows, the satellite radio market would also become a monopoly, one aligned with other big companies in the music industries like Pandora, Live Nation, and Liberty Media, complicating the notion that these new services would circumvent the negative effects of consolidation on music culture in the long run.

SATELLITES, ORBITS, AND THE STREAMING SPACE AGE

Satellite radio debuted just before Apple's iTunes Store (2003) and over a decade before Spotify became licensed and widely taken up in North

America (2011). It bridges the era of legacy broadcast media and that of the streaming media era. One thing that links these two moments and shapes our understanding of radio by satellite is the technology of the satellite itself and the broader ideas of progress, novelty, and technological anxieties that it signifies. In the space age, beginning with the launch of the Sputnik 1 satellite in 1957, a cultural fascination with outer space and new technologies defined military conflict and the boasting of power through outer space and technology, namely between the United States and the Soviet Union. And these themes have permeated popular culture from film to music. Anxieties around new technology remained fixed in place at the turn of the new millennium, with the cultural panic and uncertainty of Y2K and later with satellite surveillance and military targeting becoming a key fixture of the war on terror after the September 11 attacks. Technologies from the satellite to the internet have histories that involve their use as war machines. It's worth noting that XM's initial senior vice president of engineering was a retired air force brigadier general.[39] The satellite is an object of wonder, one that collapses space and time: it brings forth the power to see from a distance as well as to circulate music on a scale and speed that was previously not possible. In these ways, the space age remains central to how music circulates in the streaming age.

Three central concepts are used throughout this book to make sense of the influence that satellite radio has had on music culture and the circulation of music in the new millennium: orbits, perceived value, and the targeting of listeners. The orbital path of a satellite, its *orbit*, is used as a metaphor for thinking through the cycles of music consumption across the radio and music industries as well as the routines and rotations of everyday life. Songs are released and they circulate, some widely and with resonance, others with little staying power or without the result of finding an audience at the right time. Music from the past returns to the present, a fact that certainly became evident in and around the 2000s, with popular writing on the decade describing how it began with September 11, which kicked off a "single, lurid blur" that still persisted at the end of the 2010s. Amanda Mull wrote in *The Atlantic* at the end of the 2010s that it was "hard to look back at the past 20 years and argue convincingly that whatever is happening now hadn't already started in the early 2000s." Smartphones and social media have removed people from a sense of linear time, and "algorithmic timelines" mix images and sounds from any point in time. Every few years we get another *Spider-Man* remake.[40] As time moves forward, the temporal gap between past trends and cultural products that come back into style in the present is shrinking;[41] the "nostalgia loop has sped up."[42] These proclamations in the popular press, of course, are a reverberation of previous commentary and academic writing on nostalgia, pastiche, and an eternal present, debates and discussions that circulated in the 1980s and 1990s thanks to theorists like Fredric Jameson and Linda Hutcheon.[43] In early 2022 the coming together in the present of past eras in music could be explained, in part, by the prominence of catalog records in streaming music listening and by digital circulation increasing the speed of these loops. One report indicated that more than 73 percent of the American music market was claimed by catalog records, as opposed to new releases, at the start of 2022.[44]

We can think of our own tastes and preferences in music as having certain songs, artists, or albums as existing in our own orbit. They serve as references for our own interests in music and our own identities and our own sense of ourselves. Orbits reflect the cycles of music. We cycle through playlists, which often return to the beginning once they reach the end. Our fondness for a piece of music grows as we hear it over and over again. The music we grew up with might track us down later in life and in meaningful ways. Radio relates to these cycles as well, accompanying us as we rotate through the days alongside a program schedule that encourages a sense of dailiness of radio listening, a characteristic that media and communication scholar Paddy Scannell has argued to be essential to our understanding of what makes radio *radio*.[45] Orbits also shape music radio programming in referring to the time it takes for a song or an artist to be heard again on a given channel. Abrams told me that in music scheduling some orbits are tight, six hours. Others are "what we call a lunar orbit, which come up every six months."[46] Listening to SiriusXM in 2023, I regularly heard radio hosts asking listeners to contact them on social media to let them know what other channels they listen to, or what channels might be in one's orbit. I once heard the 1st Wave channel describe itself as "classic alternative orbiting the 80s."

These circulatory paths in music also speak to the ways music moves through space and time, reaching people in meaningful ways. Circulation in the realm of media and culture has been written about in the early media and communication studies work of foundational thinkers like Harold Innis and, more recently, in relation to heightened processes of globalization and the flow of cultural products.[47] Anthropologist Anna Tsing writes on interconnection and new globalism, whereby interconnection is fostered through circulation as people, money, information, and television programs circulate, and where powerful institutions spread their influence across borders in pursuit of the ongoing expansion of capital and resource extraction.[48] Media scholar Will Straw, on the "circulatory turn," argues that a turn to circulation "comes with an understanding of media as mobile forms circulating within social space." He explains that in analyzing cultural artifacts we must address the conditions that give way to cultural forms occupying social space and becoming interconnected.[49] Literary theorist Anna Kornbluh argues that immediacy in circulation is a defining feature of twenty-first-century cultural production.[50] These ideas move beyond the stages of production, distribution, and reception that have shaped much writing on media industries over the years and ask us think about moments in which these distinct categories collapse into one another and influence the more informal and immaterial movement of music.

In this moment of having access to many things all at once, the question of what gives music value, or a sense of value, when we are immersed in an expanding range of musical choice is a timely one. So what gives music value when it is cyclical? Music circulates in the present, hoping to stick with us for a period of time, and it returns easily and convincingly from the past. Some songs are lost forever, just as some satellites circle the planet as space junk; they are still out there, of course, but seemingly without much use or relevance. The satellite radio industry is one of many media industries that play a role in crafting the *perceived* value of music, which is another central concept that runs through this book. In programming decisions, radio voices, and the synergistic promotional models employed by the radio and record industries, institutions like satellite radio companies authoritatively introduce new music or select and share music from the past. When shared through intermediaries like the radio, music is given relevance and significance across time.

Writing on value, music, radio, and circulation, ethnomusicologist Timothy D. Taylor proclaims, "Word of mouth still matters, especially when it is amplified via radio. Value is created in this scene (and not only in this scene) by being on the radio, which not only might help sales of recordings and attendance at shows but also adds to a band's recognition in the scene."[51] I like this claim because it acknowledges the informal and formal circuits of music circulation, in which value is constructed and amplified. Taylor's writing is helpful for thinking through radio's influence in the circulation of music, as it allows us to consider the realization of value without the exchange of a physical object.[52] Art scholar François Brunet asks a related question in the realm of visual art and circulation: "How have pictures and objects—in different time periods—acquired meaning, worth, agency, form, even aesthetic status, by moving, or, more generally, by gaining 'currency' or 'use'?"[53] With so few large record companies and tech companies wielding influence in the music industries, it's imperative to think through the ways companies like Spotify and SiriusXM construct listening experiences and guide music fans to certain songs, concerts, albums, and so forth. The nexus of media, economics, policies, and institutions shapes our sense of what counts as *good* or as having value and plays a role in determining what music remains in wide circulation in the streaming era. As Taylor asks, "How are particular commodities valued? By whom? Why are some commodities valued more than others?"[54] In order for a music service to be successful, that is, to acquire longevity as a business and widespread application and use among listeners, it must demonstrate value to music listeners but also, increasingly, to shareholders. This creates a tension whereby music must also be devalued or strategically disposable, so that it can function across and within key policy decisions and royalty rate negotiations to ensure ongoing profitability and shareholder confidence. The inverse of this has meant an increasingly inequitable music ecosystem wherein few find success and many struggle to earn a living.

Circulation can also be heard in popular songs and the formula or structure that cycles through the verse and chorus, in which particular musical measures are repeated. Elements of a song are often cyclical, as they repeat across its duration. Critical theorist Theodor Adorno found these repetitive and standardized elements of popular song *regressive* in popular music.[55] Following the success of one popular song would be countless imitations. But meaning is made in these cycles and in repetition, and we come to identify closely and carefully with certain songs, often thanks to repetition and cyclical musical moments. In these cycles we can locate pleasure and the sort of everydayness that makes listening to music enjoyable. In Joshua Clover's celebration of Jonathan Richman and the Modern Lovers' "Roadrunner," he writes about the moment of falling in love with a song on the radio, of driving around and listening to this song on the radio, and of putting these feelings of admiration into a pop song to share with the world. It's about "going around and around," he says.[56] "Roadrunner" is not a song that pauses but instead "comes back around, a 45, a ring road, repetitive, infinite."[57] Within these pathways or routes we often encounter music after it has been accredited in some way, ascribed value and shared in a meaningful manner. This could be through a celebratory review or something casual and informal like a friend suggesting a song we haven't yet heard. The radio is just one method for encountering music in meaningful ways, but it is one with a long and powerful history, and one that is being reconstituted and recontextualized through satellite radio.

Satellite radio has been an integral component of the intertwining stories of niche and subscription listening, the siloing and targeting of taste and listening habits, and the ideal, or idealized, listening contexts of the streaming era. Simultaneously, it offers the feeling of collective connection, the long-standing domain of radio, but also of hyperpersonalization. In these ways and more, it's a radio service that is a precursor to our current moment, in which a few large music services are used by a wide variety of people to listen to their own personalized music selections. This dichotomy of centralization and fragmentation was apparent in a 2014 interview with president and chief content officer of SiriusXM Scott Greenstein, which posed this question: How do you measure success on subscription radio without advertising and ratings? In other words, how do you measure its value? His answer was that you look to subscriber numbers and growth. And what drives these numbers, according to Greenstein, is the simultaneous availability of niche music channels and styles of music and current big successful songs, which can be gleaned from YouTube's sourcing of data based on the platform's biggest music videos.[58] Beginning in 2014, the SiriusXM Hits 1 channel would regularly feature a program titled *YouTube 15*, which showcased the top 15 songs on the platform. SiriusXM, then, has simultaneously catered to mass and niche listener preferences in its approach to programming for a class of listener-subscribers. Thus, another important concept for understanding the history of the satellite radio industry is that of *targeting* listeners. The cultural technology of the satellite targets, as do, increasingly, the music companies that use listener data to shape playlists, algorithms, and channels. The satellite is a tool of military surveillance as well as one of delivering niche music channels to distinct user profiles. Satellite radio is a central figure in the platformization of music, whereby media platforms mediate the circulation of commodities.[59]

The study of satellite radio and music programming brings together technology, industry, policies, political economy, and popular culture. As such, it benefits from a media industries approach. Scholars Jennifer Holt and Alisa Perren explain that media industries research requires a range of perspectives because it involves "an extraordinary range of texts, markets, economies, artistic traditions, business models, cultural policies, technologies, regulations, and creative expression."[60] Referring to the notable influence of Adorno and Horkheimer's culture industry perspective on the study of media industries, in which a pronounced concern about the commodification of culture by media is central, they add that a media industry is not a monolith and requires a consideration of the interrelationships between media industries, their texts, audiences, and the wider societal and cultural context.[61] In Raymond Williams's groundbreaking study of television, he outlines an apt approach of analyzing the medium alongside its development, institutions, and forms and effects.[62] His approach exemplifies how media institutions create and reinforce power dynamics in society while making space for the ways that people are active in culture. In the inaugural issue of *Media Industries* journal, the editorial collective explains how the scholarly study of media industries across new conceptual approaches and methods aims to respond to the consolidation of large conglomerates; new technological modes of production, distribution, and consumption; and the sort of innovation that has sparked niche production and DIY creativity.[63] These transformations can all be understood through the history of the satellite radio industry.

In thinking through the ways that media institutions craft the perceived value of music, it's essential to emphasize the role that technologies play in shaping taste and disposition. In Charles R. Acland's study of blockbuster cinema, he writes that "a Bourdieuian analysis of dispositions toward culture needs to include dispositions toward technology, format, and platform."[64] And in the case of satellite radio, dispositions toward technology also include dispositions toward subscriptions, access, mobility, and the feeling of being able to navigate an ever-expanding lineup of music channels. Acland refers to Pierre Bourdieu, whose influential writing on hierarchies in society and culture has been widely cited in writing on taste, distinction, distaste, and class, often in reference to music and its central role in the formation of identity and one's orientation in the world. In our subscription-laden media landscape we face even more choice, and the role of class, as well as the discerning force of taste, becomes crystal clear in tiered subscription plans.

The technology of the satellite also urges a careful consideration of the way that tropes of newness permeated the launch of the new radio services in the early 2000s. Satellite radio has a complex relationship to time, caught between rhetoric of newness and the longer history of radio. Writing on the aesthetics of media in transition, David Thorburn and Henry Jenkins argue that "the process of media transition is always a mix of tradition and innovation, always declaring for evolution, not revolution."[65] As media historian Lisa Gitelman explains, the introduction of new media is never entirely revolutionary.[66] We must be wary of revolutionary claims of newness, especially when they are instigated by the industries and institutions driving the development of technology and policy.[67] Hearing satellite radio through a framework of media in transition accentuates music programming and listening experiences that amplify radio's cultural past, both connecting it to and distinguishing it from new media. With satellite radio, a discourse of newness and revolution drives interest in the new services and in the act of committing to a subscription, but the resources and institutional weight of legacy media, namely radio, help to sustain the service across tumultuous years of change in music listening and consumer technologies. There is then a sense of both hybridity and fluidity in our understanding of satellite radio. The hybridity of the technology, that *in-betweenness* of radio, satellites, and internet, encourages a fluidity across programming. Programs come and go. Some channels are paused for temporary "pop-up" channels or removed for new permanent channels, and guest hosts regularly take over the airwaves. Satellite radio simultaneously breaks new songs while leaning heavily on back catalogs and artist legacies. Hybridity and fluidity have enabled SiriusXM to continually increase profits and subscribers during a tumultuous transition to digital and streaming listening contexts.

A theme of hybridity also suggests that so-called new media must be understood through a historical lens that accounts for these ties to a wide variety of industries and their own respective histories. A discourse of *new* that surrounds a new product or service, like satellite radio, serves to convince consumers of its superiority, framed against that which can be delineated as old or outdated. I visited the SiriusXM studios in Washington in 2017; during my visit, an overall sense of newness and distinctiveness from how one might imagine a broadcast radio studio was apparent throughout the sprawling and open warehouse located in Washington's NoMa area, a spot that has been gentrified since the time of XM's establishment. There were rows of smaller broadcast booths and then some larger ones where talk shows, often with multiple hosts and guests, could interact. A space with large servers and cooling fans and two large satellite dishes were visible through a window. The space was very open and was once a factory for printing *National Geographic* magazines. My tour guide explained that the early days were quite "hippyish, frontierish," people acting on a wild idea that seemed to be new and revolutionary at the time. Hybridity and fluidity speak to the overall organization of music programming on the satellite radio channel lineup, in which eras and musical styles are juxtaposed and aspects of broadcast radio collide with initiatives and experiments driven by attempts to carve out space in emergent media forms and programming trends. But hybridity also speaks to the industrial organization of satellite radio and the centrality of partnerships with existing industries like the automobile and consumer electronics industries. As soon as I entered the building, and before passing through the security desk, a large poster greeted me that read: "Most Wanted by SiriusXM: 85% of dealerships are participating in our pre-owned program giving subscribers free 3 month trial subscriptions. Here are some of the few that aren't. Know someone? Think you can help?" This resembles a sort of *combining mecha* approach, whereby key businesses, devices, and products come together to form satellite radio (not unlike the monstrous machines of Japanese science fiction, anime, and manga). Satellite radio is nothing without the intermingling of satellite technologies, the radio industry, and the automobile industry.

Industry partnerships had significant material consequences in the early days of XM and Sirius, and they still did at the time of writing. XM explained that the company was funded through its planned commencement of commercial operations in the summer of 2001 and into 2002 with the help of "strategic and financial" investors that included General Motors, Clear Channel Communications, DirecTV, Columbia Capital, American Honda, Telcom Ventures, and others. The company had contracts with Delphi-Delco Electronics, Sony, Motorola, Pioneer, Alpine, Mitsubishi, Audiovox, Clarion, Sharp, Blaupunkt, Fujitsu Ten, Hyundai Autonet, Bontec, Visteon, Panasonic, and Sanyo to manufacture and distribute XM radios.[68] Companies including Circuit City, Best Buy, Radio Shack, Sears, Mobile-One, and Magnolia Hi-Fi were on board to distribute and promote the new products and services. On the technology side, Hughes Space and Communications, Hughes Network Systems, Alcatel, Telesat, LCC International, Lucent, STMicroelectronics, and the Fraunhofer Institute were partners with XM.[69] Even in SiriusXM's acquisition of Pandora in 2019, the company proclaimed that it was continuing its established position as a leader in audio entertainment in North America, in part, thanks to "deep and long-term commitments from automakers."[70] Competing audio media companies like Spotify would later look to collaborate with automakers, establishing connections with Uber, BMW, and MINI.[71]

Partnerships with industries with long histories complicate the revolutionary claims of newness that have been mapped onto satellite radio, and so too does the technology of the satellite itself. Satellites lent music services an essence of the *new* in advance of the launch of Sirius and XM, but they were integral to commercial radio broadcasting for decades before that. Even the prepackaged streaming music service, Muzak, used satellites for distribution, with all 218 Muzak distributors in the US transitioning to satellite service in 1980, having previously used taped programs sent out from Muzak's headquarters. The new satellite distribution was said to increase the flexibility of programming and upgrade sound quality.[72] Radio scholars Alexander Russo and Bill Kirkpatrick have noted that even before the Telecommunications Act of 1996 there had been a longer sense of standardization in the corporate music industry and in American life more broadly, and that there had been a much longer and significant role of the satellite in radio history and practice.[73] In the early 1980s, satellites were said to make radio an "immediate medium" and to "help promote radio as a national, rather than local, medium."[74] The American public radio network NPR had 192 stations linked together by satellite in 1980, which would allow for a "high fidelity stereo" sound for music radio shows like *Jazz Alive*.[75] With satellite technology, NPR could bring forth classical music programming of a quality of transmission that would add "an impression of space so that you can feel the four walls of the room [they're] broadcasting from." NPR senior classical music producer Fred Calland said, "When we say we're taking you to a concert hall in Salzburg, you'll get an aural sense of actually being there."[76] In these ways, the technology of the satellite has shaped perceptions about the quality and immediacy of radio broadcasting.

SATELLITE MUSIC RADIO

In 2016, one analyst said that SiriusXM "probably isn't the first name that comes to mind when you think about subscription content, yet it's one of the most successful companies in that category."[77] SiriusXM, and its organization and circulation of music programming, developed a subscription model for music listening, the very business model that has become so common, even dominant, in music listening in the twenty-first century. In 2019 *Billboard* proclaimed, "Each music listener is worth more than ever." In the two years before that, SiriusXM had gained a 1 percent share of listening time, adding 1.1 million subscribers worth $250 million in annual revenue. Across the board, an increase of 29.6 million American music subscribers boosted total industry revenue by 22.4 percent over two years, all from the same number of music listeners and listening hours, more or less.[78] Subscription satellite radio has amplified the role of celebrity talent on the airwaves, with the acquisition of big names to grow subscribers. Between 2006, when Howard Stern moved to Sirius from terrestrial radio, to early 2014, Stern earned more than $700 million, with some speculating that he has made more than $2,000 per broadcast minute.[79] As broadcast historian Michele Hilmes writes, internet, satellite, and digital radio "are not so much replacing but *restaging* radio as a global medium."[80] Radio, as it enters this transnational era, according to Hilmes "is once more at the cutting edge." Over SiriusXM's history, the company has been at the center of this restaging of radio as well as music.

The focus of this book is music on satellite radio. But satellite radio is also famous for its talk programming. News, sports, and entertainment have all been major draws for subscribers. From Howard Stern's consecutive Sirius contracts to the acquisition of podcasts, the ongoing significance of talk programming to SiriusXM's business model is obvious. But my primary interest is in the connections between the radio and record industries by way of satellite radio, as well as the place of satellite radio within the realm of subscription music listening. Others would be wise to pursue questions about how talk radio functions on SiriusXM, but that is beyond the scope of this work. Thus, this book is situated in a field of research that lies at the intersection of radio studies, popular music studies, and media industries studies. In radio studies, music radio is arguably comparatively understudied as compared to the cumulative range of topics affiliated with talk radio: radio drama; news reporting; talk-based educational programming; and especially in recent years, the growth in research on podcasting, an area within which talk-based audio programming dominates. That said, the study of music radio is indeed growing and has been the focus of a range of compelling articles and books.[81] Music has been, and continues to be, a central presence in radio's social and cultural fabric.

The story of satellite radio is also one of media monopolies and industry concentration, given the merger between Sirius and XM and the growing power of Liberty Media in the music ecosystem. Particularly since the 2020s, Ticketmaster and Live Nation have become the targets of anti-monopoly sentiment in policy circles and in music fan communities as unfair business practices have increased the barriers to access in the live music sector and beyond.[82] Liberty's chairman and largest voting shareholder, billionaire John Malone, has worked in cable and telecommunications with companies like Bell Telephone Laboratories and Tele-Communications, Inc. (TCI). He is also responsible for the term "500-channel universe" and is the second largest landowner in the United States. Liberty Media holds a 35 percent stake in Live Nation, and its subsidiaries include the Formula One Group and the Atlanta Braves. The history of Sirius and XM is also a story of Liberty Media's increasing corporate influence on music in North America and beyond. SiriusXM has gradually embraced a more automated and less adventurous model of programming as it becomes shaped by streaming media and music platforms and the ongoing drive to acquire more subscribers. As Eric Drott argues in *Streaming Music, Streaming Capital*, a "notable feature of streaming platforms is the way they decouple music's use value(s) from its exchange value. Services pay for music, while users pay for the service."[83] SiriusXM serves as an example of a media company that has employed a variety of programming practices to court subscribers over the turn to streaming listening and has continually worked and reworked the ways it ascribes value to music in order to accomplish this feat. Satellite radio is both reflective of radio's resilience and ongoing cultural significance and a key example of how industry concentration, platform capitalism, media monopolies, and financialization are putting the vibrancy of diversity of music culture in jeopardy.[84]

The methods used in this book are strategic and reflective of the challenges that come with researching a closed-off, publicly traded corporate media company like SiriusXM. I've used document analysis and close reading, particularly in accumulating and carefully reading annual reports, proxy statements, and press releases. In Acland's *American Blockbuster*, he explains that reading, "targeted, focused, and purposeful," is a helpful method for sorting through trade articles and for becoming accustomed to the stories that emerge when one spends time with these sources (as opposed to machine reading, for instance).[85] Reading enables the argument to emerge, after the searches have been done and the results have been organized. Acland adds, "The hard work of assembling material, locating additional sources not covered in the digital corpus, figuring out the contextual matters, and beginning to construct an argument that will illuminate and bring understanding still has to take place" after search results are returned.[86] With some sources, like annual reports and press releases, I have been careful to account for the rhetoric of optimism and strategic maneuvering that these documents tend to employ, as they aim to assure shareholders of future success and the capacity to weather competition and wider industry turmoil. But within these documents one can ascertain broader shifts and trajectories and get a sense of how a media company is situated in the broader media industries and how the company imagines its future aims and priorities. They also include helpful financial statements that show where the money has gone and is going. I have read through an exhaustive amount of trade press coverage, namely in publications like *Billboard*, *Variety*, and *Radio & Records*, and carried out analyses of select satellite radio shows and archives of channel lineups and descriptions. Some of this material has been found in annual reports, some on collaborative wikis online, some through navigating websites like the Internet Archive, and some through just listening to the radio regularly and for long periods of time. Over the years since Sirius and XM launched, *Billboard* coverage on these companies has been particularly extensive. However, over time this extensive coverage has begun to wane, likely due to staff and resource reductions and the shift from print to digital. Ownership of *Billboard* moved from VNU Media/Nielsen to e5 Global Media Holdings to the Hollywood Reporter-Billboard Media Group, under the holding company Eldridge Industries.[87] *Radio & Records* also covered Sirius and XM until its final issue in 2009. The publication was independent until *Billboard's* parent company at the time, VNU Media, took it over in 2006. Other news outlets, like *RAIN News*, a company with a daily digital industry digest, became more prominent over the SiriusXM years, with extensive coverage of Pandora, podcasting, and SiriusXM.

Part of the challenge in researching a media company like SiriusXM is that there are barriers to accessing workers to conduct interviews or to locate and read internal documents that have not been made public. This is not unlike how some researchers have described the study of Spotify, a "macroactor" with power that relies on the ability to keep certain key aspects of the company, like a recommendation algorithm, "firmly closed" and convince us that it could not be otherwise.[88] In researching SiriusXM, I had difficulty speaking with workers about their experiences at the company, aside from one radio DJ, whose helpful insights appear sporadically throughout the book, and a few former employees, including XM co-founder and media executive Lee Abrams. The coming together of these methods is designed to generate an understanding of the institution within its wider musical and cultural context and is part of a research trajectory of understanding how institutions become *everyday* and shape our ongoing relationships to music.

Each of this book's chapters deals with a section of the historical time line of Sirius, XM, and SiriusXM, which mostly flows in chronological order. Each chapter deals with its own sets of issues that connect to broader questions and themes in popular music culture and media. The first chapter is about the early years of Sirius and XM and how the cultural technology of the satellite shaped ideas about newness and value in music and radio programming. It draws on the metaphor of orbits to inform an understanding of satellite radio programming and of popular music culture in the new millennium. The next chapter stays within this historical moment and gives attention to the strategies used by Sirius and XM to develop and grow their respective subscriber bases. It asks: How did Sirius and XM work to transform radio listeners into music radio subscribers? Given the comfortable relationship between popular music and material objects of the 1990s, when CDs reigned as a popular music format, many wondered whether the impetus for people to own things would prevent subscription models from being widely taken up. Sirius and XM would need to convince listeners that radio was worth paying for. In chapter 3, music programming and the channel lineups of Sirius, XM, and SiriusXM are discussed, particularly the role of music in crafting a sense of perceived value via prestige radio. This chapter describes how the satellite radio companies have drawn on nostalgia, music heritage and tourism, niche fan cultures, and sounds that have been neglected by broadcast radio to use music and its meaningful qualities to grow and maintain their base of subscriber-listeners.

In chapter 4 the time line centers on the years around the merger of Sirius and XM in 2008 and the policymaking and processes of royalty rate negotiation that placed music in a paradox: simultaneously, they served to demonstrate the perceived value of the subscription services while also strategically framing music as less valuable or as disposable to the overall business model so as to circumvent higher royalty payments to musicians. This chapter shows how the history of SiriusXM is one of corporate consolidation and of radio and record industry power that often sets the terms of how musicians work and live. Chapter 5 flips back to the discussion of the perceived value of music on satellite radio by showcasing key programs and radio hosts who exemplify the sort of music radio programming that Sirius and XM have developed. The expansion of SiriusXM into

Canada is discussed in chapter 6. It highlights the shift in spatial and mobile listening practices for satellite radio and other music services, from the space of the automobile to a wider variety of mobile and private listening modes facilitated by smartphones and other mobile devices. Finally, chapter 7 deals with the more recent years of SiriusXM, as of this writing, including the acquisition of Pandora, the influence of platforms on music and culture, and the growth in ownership and control of Liberty Media, and ponders the future of this increasingly significant player in the music and media industries going forward. Let's launch into it.

1 The Songs Down to Earth

MUSIC RADIO AND OUTER SPACE

When satellite radio launched in America, the top 5 *Billboard* 200 albums came from a range of music genres, with Jay-Z, Nickelback, Alicia Keys, Fabolous, and Bob Dylan occupying the top spots. This was also a unique time for less-than-mainstream bands. Many were signing deals with major record labels, which hoped to strike gold with the next rags-to-riches success story.[1] Satellite radio, with its variety of music channels, was well equipped to fold a variety of streams of music into its service at the turn of the millennium, reaching listeners nationally and across taste categories. But not every song or artist could become a regular feature of the music heard on XM Radio and Sirius Satellite Radio. As subscription radio services, they gave careful consideration to the crafting of music programming in order to demonstrate an exceptional listening experience by comparison to commercial broadcast radio. Satellites, as cultural technologies linked to ideas about the future and the endless expansiveness of outer space, construct a notion of value for music on satellite radio.

This chapter investigates the connections between outer space and satellites, music and radio. Outer space, and its myriad knowns and unknowns, has shaped popular music history in fascinating ways. Outer space presents music as otherworldly, experimental, or exceptional. The satellite is a technology that encourages this mythology. It's an object beyond our reach but one that we sometimes catch a glimpse of as it moves across space. As XM and Sirius launched their satellite radio services, in 2001 and 2002 respectively, the figure of the satellite helped convince listeners of satellite radio's superiority to terrestrial radio. One way that these new services were distinguished from terrestrial radio was through an association with outer space. Throughout the early years of Sirius and XM, a number of values were attributed to satellites, and they shaped ideas about premium or prestige radio. The satellite reinforces notions of authenticity and masculinity in histories of technology and musical genres; it draws on our relationship to space and geography to communicate ideas of expanded radio coverage within, but not beyond, North America; and it affects our sense of time in the way that technologies and media industries centralize and standardize the delivery of music and radio programming. Satellites orbit Earth, and the notion of the orbit shapes satellite radio programming as well as our experience of popular music. Music circulates and enters and exits our everyday lives, processes facilitated by major media institutions like the satellite radio industry. I return to the concept of orbits at the end of this chapter.

RADIO INTO SPACE

As an object that shapes media, culture, and state power, the satellite figures prominently in the postwar era. It is bound to an era of increasing globalization of culture and communication and signifies aspirations to the instantaneous sending of data and information by new technologies and tools developed during the space age. Science fiction writer Arthur C. Clarke said that his article in *Wireless World* in 1945 was "almost certainly the first appearance of communication satellites in any newspaper in the world."[2] Clarke proposed an idea of "manned space stations as a means for communicating between various points on the earth's surface, and for broadcasting to areas of the earth."[3] His vision involved synchronous satellites in orbits 22,300 miles above the earth.[4] It would take some time before aspects of this vision would be put into practice, as *unmanned* satellites became a crucial component of how people around the world communicate and conduct their daily lives. Early satellite communication services were for basic telephone and telegraph services and based on an analog 4 kHz telephone channel. This channel could be combined with others to provide a wider-band channel for broadcasting or data.[5] Lower entry costs and increased access to satellite technologies helped them become a more routine and present aspect of our lives, especially through television in the 1970s and 1980s.[6] By the end of 1986 there were nearly 100 telecommunications satellites transmitting to Earth from the geosynchronous orbit,[7] and there was a total of 538 satellites in this orbit as of 2021.[8] Though not always visible, their function and influence are now undeniable.

Satellite CD Radio Inc., the former company name of Sirius Satellite Radio, was founded in 1990 in Washington, D.C., by Martine Rothblatt, David Margolese, and Robert Briskman. Rothblatt had helped launch PanAmSat, a company providing commercial satellite bandwidth for broadcasters.[9] Satellite CD Radio was the first to petition the FCC to assign unused frequencies for broadcast use by satellite, a move that the radio industry vowed to block "at every turn" out of fear of increased competition and perceived obsolescence by comparison.[10] Rothblatt resigned as chair and CEO in 1992 to pursue a medical research foundation focused on curing her daughter's illness, and former NASA engineer Robert Briskman, who had designed the core technology for satellite radio, stepped in as replacement.[11] Half a year later, David Margolese, the co-founder of Rogers Wireless in Canada, succeeded Briskman. Under Margolese, the company was renamed CD Radio, which in 1999 would be changed again to Sirius Satellite Radio to avoid association with the soon-to-be-outdated CD technology. Sirius received a financial boost in November 1998 when it inked a $200 million agreement with Apollo Management investment firm. This followed earlier stock purchases of $100 million by the investment group Prime 66 Partners.[12] Before Sirius officially launched in July 2002, it received a $150 million loan from Lehman Brothers after passing a series of tests, including having its signal checked in wooded rural areas.[13]

XM Satellite Radio, the company that would be Sirius's competitor for a number of years before the two merged in 2008, launched in San Diego and Dallas-Fort Worth on September 25, 2001, to the sound of Bob Marley's "One Love." Then president of XM Hugh Panero proclaimed that the new service, the first to begin operation in the US, was "the signal of the future," a concept that was said to be "part rocket science, part rock 'n' roll." XM expanded nationwide on November 12, 2001.[14] The company proudly acknowledged this feat in its 2001 annual report, complete with an image of a mother, daughter, and two young men staring into the sky, basking in the glow of what we can assume to be the launch of a satellite by rocket.[15] Before taking on the brand name XM Satellite Radio in 1997, the company began as a division of the American Mobile Satellite Corporation: the American Mobile Radio Corp. American Mobile Satellite was founded in 1988, and the division was created in 1992. WorldSpace, a company with satellite radio service for Africa and the Middle East, and also the first to launch a comprehensive schedule in 1990,[16] became an investor in the company in 1992; in 1997, XM became the second company to obtain a satellite digital audio radio license from the FCC. At one point Sirius, then named CD Radio Inc., filed a patent infringement suit against XM, claiming that its proposed transmission service was infringing on three CD Radio patents, and that XM should be prevented from constructing its system.[17] However, the two companies eventually decided to share various aspects of their technology.[18] On July 10, 2000, XM boasted news of a recent fundraising campaign that had generated $235 million, and the company was now "financially fit to fly."[19] This followed a round of investment from companies like General Motors Corp., DirecTV Inc., and Clear Channel Communications, which totaled $250 million.[20] The investment from Clear Channel was surprising, given concerns from broadcasters that the new satellite services were bringing forth unprecedented competition, but the radio station operator was providing programming to XM and looking to benefit from the new service.

Before radio services could begin, each company had to launch its own satellites. Sirius launched Sirius-1, the first of its Loral-produced, three-satellite constellation, from the Baikonur Cosmodrome in Kazakhstan on June 30, 2000, the same location from which Sputnik had been launched in 1957. This followed delays and earlier mishaps due to concerns about the manufacturing of Proton rocket second-stage devices built before 1993; Sirius spokesperson Terrence Sweeney explained that these issues had been corrected, and that second-stage devices built after 1997 were being used in the launch.[21] The remaining satellites were scheduled

Figure 3. Watching a "satellite launch." *Source:* XM Satellite Radio, *2001 Annual Report*, 2002.

for launch by the end of the year. Margolese said that the first satellite would allow the company to verify the capabilities of its system, including a coast-to-coast signal and digital-quality sound.[22] Sirius noted that the construction of satellites cost nearly $129 million during 2001, with the total three-satellite system costing just under $945 million.[23] The three satellites were named Radiosat 1 through 3 because a fleet of satellites was already named Sirius, but they became commonly referred to as FM-1 to FM-3. Two were in rotation over the United States, and one was parked over the Caribbean.[24] The three orbiting satellites were described by Sirius as traveling in a figure eight pattern that extended above and below the equator, spending around sixteen hours per day north of the equator. At all times, two of the satellites were north of the equator, with the third traversing south while not transmitting. This configuration was said to "yield high signal elevation angles, reducing service interruptions from signal blockage."[25]

By 2006, XM had launched four satellites: XM-1 in March 2001, XM-2 in May 2001, XM-3 in February 2005, and XM-4 in October 2006. XM-3 and XM-4 replaced the first two XM satellites and were deployed in geostationary orbits at longitude 85° west and longitude 115° west.[26] A key difference between the initial Sirius system and that of XM (and Sirius's later satellites) was that the former would travel in elliptical orbits at 23,000 miles above Earth, while XM's pair of more powerful satellites were geostationary at approximately 22,300 miles, or 35,785 km, above Earth, which is the altitude for satellites in the geostationary/geosynchronous orbit. The distinction between the terms *geostationary* and *geosynchronous* is that the former is at zero inclination and sits above the equator. Both match Earth's rotation period; they are in sync. The geosynchronous equatorial orbit (GEO) is a thin gravitational ring that is a few kilometers thick and wide; it is central not only to global communications, but also military and intelligence agencies. By contrast, the low earth orbit (LEO) is a region about 160 to 2,000 km from the earth, and this is where the majority of Earth's satellites are based, primarily those used for remote sensing and imaging, scientific monitoring, and phone networks.[27] If you happen to be somewhere very dark and spot a satellite in the sky, it would be close to the earth in the LEO.

The first two XM satellites were said to be some of the most powerful ordered at that time, with each providing 18 kilowatts of total power. In order to generate such high power, two solar wings each use five panels of "high-efficiency, dual-junction gallium arsenide solar cells." The satellites also carry "the flight-proven xenon ion propulsion system (XIPS) for all on-orbit maneuvering."[28] After the third and fourth satellites were in operation, XM-1 and XM-2 began to function as in-orbit spares.[29] With plans to construct a fifth, XM had spent approximately $869.8 million on manufacturing, launch costs, financing, and in-orbit performance incentives, with additional costs for collocation.[30] The expected lives of these satellites are approximately fifteen years, with any decreases in this amount of time due to issues like defects in construction, electrostatic storms, collisions with other space objects, and other events like nuclear detonations occurring in space.[31]

Well after the merger of the two competing companies, SiriusXM outlined the components of its fleet of orbiting geostationary satellites as of 2021: FM-5 and FM-6 transmit the service on frequencies originally licensed to Sirius, and XM-3 and XM-4 transmit on frequencies licensed to XM. XM-5 serves as a spare for both systems in case of a failure of any of the other satellites.[32] A few years earlier, in 2016, the Sirius network transitioned to the geostationary orbit system from its system of highly inclined elliptical orbits. This meant that FM-1, FM-2, and FM-3 were moved into disposal orbits.[33] On December 13, 2020, SXM-7 was launched for in-orbit testing, which resulted in failures of certain payload units.[34] The company also entered into an agreement for the design, construction, and launch of SXM-8. Both SXM-7 and SXM-8 were intended as replacements for XM-3 and XM-4, indicating an ongoing process of maintaining the fleet of orbiting satellites. Evidently, as this brief overview of SiriusXM's fleet of satellites demonstrates, establishing and sustaining a satellite radio operation is not cheap and involves startup and maintenance costs as well as insurance considerations and the ability to navigate regulations and policies at national and global levels.

With satellites in orbit, the two companies were also establishing a footprint in two of America's most prominent cities. In New York, CD Radio built a $50 million and 100,000-square-foot facility in Rockefeller Center.[35] The facility is housed in the 36th and 37th floors of the McGraw-Hill Building, now known as 1221 Avenue of the Americas, and from the start, it was equipped with 4.2 terabytes of digital audio storage (enough for ten

Figure 4. SiriusXM's SXM-7 SpaceX satellite. *Source:* SiriusXM press release, December 13, 2020. *Credit:* SpaceX.

Figure 5. SiriusXM's SXM-7 satellite in orbit. *Source:* SiriusXM press release, December 13, 2020. *Credit:* Maxar Space Systems.

thousand hours), 112 output channels (50 channels for commercial-free music, 50 for talk programming, and 12 for backup), and sixty-three workstations. Bethany McLean described the new Sirius studios for *Fortune*: "The $38 million space, which Margolese designed, is all ultramodern hipness: clean blond wood, stainless steel, spotless white walls, and planes of glass."[36] Songs could be extracted from CDs and stored as MPEG-2 files in a bank of servers. Sirius program streams, with an average data rate of 384 kilobits per second, were then fed into a Lucent multiplexer before uplink from Vernon Valley, New Jersey.[37]

XM Satellite Radio built its headquarters and broadcast center in Washington, D.C., a city with a storied music history of its own, which has been dubbed "The Netplex" due to the number of high-tech companies it was home to.[38] XM's home base, a former Judd & Detweiler printing plant in the city's northeast, is a 150,000-square-foot facility with eighty-two digitally interconnected studios, the largest audio broadcast facility of its kind in the US.[39] Suspended studios were said to provide "such superior acoustic isolation that classical music and chainsaw rock 'n' roll can be played side by side without compromising either's sound quality."[40] The independent suspended studios had four-inch-thick, steel-clad structural panels to seal out intrusive noise, for complete acoustic isolation. Studio innovations included a large-scale automation system, with Encoda System's Paradigm suite managing all the audio channels and satellite uplink parameters.[41] Two on-site, seven-meter satellite uplink dishes could transmit one hundred channels to the satellites.[42] In 2002, audio data were stored on four hundred workstations that could hold 1.5 million songs, or fifty terabytes of data, more than four times what was held in the Library of Congress.[43] The company also leased technical space in Florida, studios and offices in New York, and space for a listener care center in Virginia.[44] Studios were also opened in Nashville at the Country Music Hall of Fame. Years later, the merged company based its corporate headquarters in New York City, with both the New York and Washington studios originating programming and additional programming generated in smaller studios in Los Angeles, Nashville, and smaller venues across the country that have included Cleveland, Memphis, and Orlando.

In addition to satellites and studios, the automobile is an especially integral component of the overall satellite radio infrastructure and business model. Key alliances with major American automobile manufacturers were apparent before the companies launched their respective services. General Motors had a 25 percent interest in XM through one of its subsidiaries, Hughes Electronics, a company that built the XM satellites.[45] Of course, GM would then sell vehicles equipped with XM radio. XM, with a brief and radio-friendly brand name, envisioned its place on the dashboard as the third band on any standard receiver: AM, FM, and XM. Early reports on Sirius's plans anticipated that Ford vehicles would receive and decode the MPEG-2 compressed signals.[46] The vehicle was also a space in which the audio quality of satellite radio could be imagined, given the isolated sonic space of the vehicle's interior. In fact, a study of American consumers conducted just before XM began broadcasting

Figure 6. XM Studios, Washington, D.C. Photo by author.

found that 82 percent wanted surround-sound audio as opposed to stereo in their cars.[47] A digital signal for radio would enable the use of 5.1 audio with multichannel sound, something that FM signals were incapable of doing.[48] XM's annual report for 2005 announced that it would begin broadcasting select channels in 5.1 surround sound, twenty-four hours a day beginning in early 2006.[49]

Establishing audio dominance on the dashboard was not a given, however. Other viable entertainment choices in addition to AM and FM radio were available to drivers, such as cassettes, CDs, DVDs, MP3s, and the impending arrival of wireless internet in the vehicle. This had some consumers and tech writers questioning whether satellite radio would become the next Betamax: a failure. As of summer 2003, most satellite radio consumers purchased a receiver from an electronics store for aftermarket installation. The move to have receivers available as a factory-installed option was crucial for growing subscriber numbers. Both Honda and

GM, two of XM's largest shareholders, planned to add factory-installed models in select 2003–2004 vehicles, and Sirius established deals with DaimlerChrysler, Ford, BMW, and others.[50] FM-modulated radios as well as three-band radios (AM/FM/SAT) would rely on a set of integrated circuits (a chip set), enabling the device to decode, decompress, and output the broadcasts.[51] Signals reached the receivers from the satellites, although some regions were unable to receive clear signals due to interference from tall buildings, bridges, and other obstacles. Because of this, both Sirius and XM have relied on networks of proprietary terrestrial repeaters to improve signal quality.[52] The satellite radio system is one of many integral components.

LICENSING SATELLITE RADIO

To bring satellite radio to consumers, Sirius and XM also required licenses. As with terrestrial radio, satellite radio required FCC permission to operate in the United States. As early as 1959, the FCC was imagining that communication with objects in outer space meant that radio was no longer terrestrial and that an expansion in usable parts of the spectrum would help accommodate a broader use of radio.[53] There had been concerns in the wake of Russia's launch of Sputnik that satellites would "jam spectrum," with speculations about a space age that could bring a "swelling demand for radio frequencies for telemetering, for guidance and controls and, for the day of human space travel, communications."[54] Given the global nature of satellite communications and broadcasting, international governing bodies are also involved in regulating satellite communications. All radio frequencies are required to be registered with the International Telecommunications Union's (ITU's) International Radio Frequency Registration Board, and this applies to those frequencies used in the geosynchronous satellite orbit.[55] A 1963 Conference on Space Communication held by the ITU in Geneva, Switzerland, resulted in the allocation of 2,800 mc of spectrum for space communication satellites.[56] This conference also sought to ensure that satellite transmissions from the GEO did not interfere with other signals on Earth or with transmissions from other satellites.[57]

In 1992 the FCC allocated the portion of the S-band (2.3 GHz) between 2320 MHz and 2345 MHz for exclusive use by SDARS. XM was using 12.5 MHz of this bandwidth for signal transmission, from 2332.5 to 2345.0 MHz, as was Sirius within the 2320.0 to 2332.5 MHz frequency.[58] XM's bandwidth was subdivided to carry six signals: two from each of XM's two satellites and two by terrestrial repeaters for signal reinforcement. XM's radio content, both data and audio, was represented by two carriers, with the other four carrying duplicates of the same content for redundancy. By contrast, the Sirius receivers would decode three 4 MHz carrier signals at once. A receiver combined audio from multiple channels from a signal transmitted in Time Division Multiplex format.[59] In 1992 the FCC proposed reserving enough bandwidth to allow several services to operate, but at this time, only Satellite CD Radio Inc. had applied for a license. Five years later in 1997, the FCC "set aside a portion of the airwaves" for satellite radio services, and this led to four companies bidding on two available licenses to provide services. In addition to CD Radio and American Mobile Satellite Corp., the other two companies that requested the FCC take action in this area were Digital Satellite Broadcasting in Seattle and Primosphere in New York.[60] At the time of application to the FCC, CD Radio was informed that it needed to raise nearly $200 million before receiving final FCC approval. The XM and Sirius applications took seven years for the FCC to approve, mainly because the National Association of Broadcasters (NAB) charged that "the new service threatened 'traditional American values of community cohesion and local identity'" as well as its members' revenues.[61]

Although Sirius and XM were thought of as being in the radio business, the licensing process outlined ways that the new services were distinct. Representatives of Sirius explained in 2003 that the company was regulated by the FCC as a subscription-based, non-common carrier, not a broadcast service.[62] The distinction against broadcast services meant that the company was not bound by foreign ownership provisions of the Communications Act, and as a private carrier, it was free to set its own prices and serve customers without economic regulation.[63] Another way that the satellite services were regulated as distinct from terrestrial broadcasters was through FCC rules that required XM and Sirius to exclusively provide national programming and not compete locally.

A further unique characteristic of satellite radio, specifically in the post-merger context, is the fact that one company, and one company only, delivers a satellite radio service. The FCC grants permission to launch and operate subsequent satellites, and for many reasons it's not an easy feat to build and license a satellite service. In regard to Sirius and XM launching in its own spectrum, Margolese said, "We own this spectrum." This quote has been described as indicating that he believes he "solved the problem that has always bedeviled the media business, which is competition."[64] Evidently, Margolese's triumph of owning the spectrum would become an essential component of SiriusXM's longevity in the media industries. It exemplifies monopoly power and its control of key communication infrastructure that ultimately shapes the circulation of music.

THE SATELLITE AS CULTURAL TECHNOLOGY

Satellites are indispensable to the operation of SiriusXM (and previously, both Sirius and XM), and they require significant financial investments. And there are broader cultural and musical meanings that are mapped onto the cultural technology of the satellite. A *cultural technology*, according to communication theorist Jody Berland, connects the "various processes and practices that comprise culture." As Berland expands, "It refers to the formal, phenomenological, and social properties of media technologies together with the machineries of knowledge and power through which they emerge and within which they work, and it acknowledges the subjects and subjectivities produced through interaction with these technologies along with their heterogeneity and ambivalence."[65] Cultural technologies like the radio and the satellite mediate space and mediate between the production of music and the production of audiences. Technologies are not neutral in their formation or use, and they affect our social lives in myriad ways.

The satellite provides a material, technical, and conceptual layer to our understanding of what radio is or could be, one largely advanced by the marketing and advertising of these services. Sound, particularly music, was said to be of enhanced quality or fidelity by virtue of the satellite. One report from *CMJ New Music Report* indicated that Sirius and XM were competing for the best quality sound in satellite radio. At one point, XM was "boasting about the results of a recent side-by-side audio quality test" that deemed it superior because of its CT-aacPlus audio encoding with Neural Audio optimizing.[66] A few weeks later, Sirius announced its use of a new sound-enhancing technology, the PAC v4 Audio Codec.[67] That said, the actual quality of the sound of satellite radio has regularly been critiqued as not matching the promise of CD quality sound. XM, for instance, explained that its music channels require about 56 to 64 kilobits per second (kbps), and talk channels require less bandwidth.[68] But subscribers have not always been convinced of its quality, and as the companies grew, they worked to balance the number of channels against each one's quality. At 64 kbps, with the AAC encoding format, the quality is fairly standard and not exceptional. A more tangible example of improved sound quality is the fact that the signal does not fade or become noisy with static, as does terrestrial radio. Regardless, the figure of the satellite adds value to our attitudes about radio, values that are often perceived and constructed through the intermingling of music, subscription dollars, and assumptions about the power of communication technologies. The allure of the satellite and its ability to beam down music from space obscure the murkier questions about the overall quality of the signal's sound. This makes sense given the satellite's longer history of bringing improved sound to radio. For example, *Billboard* reported on a satellite transmission from the San Francisco Opera House on Chicago's WFMT-FM in November 1978 and called this "the first use of high quality stereo transmission by satellite for a live radio-only broadcast."[69] When American NPR stations began to make use of satellite interconnection to replace terrestrial linkage in 1979, the improvement in sound quality was routinely highlighted, particularly in a shift from monaural sound to stereo.[70] The satellite mediates space, as has music, and this strange and expansive zone imbues a level of prestige, or higher value, to everyday cultural products.

    The satellite is essential infrastructure for the circulation of music radio and, more broadly, everyday communications and cultural exchange. Media scholars Lisa Parks and Nicole Starosielski call attention to the role of "signal traffic" in helping us understand how the content we watch, listen to, and read moves through the world and how this movement shapes and affects its form. A key component of this movement, or traffic, then, is infrastructure. Sites and objects like cellphone and broadcast towers, undersea cables, and of course satellites, all support the circulation of audiovisual signal traffic.[71] Infrastructure "scaffolds both social and environmental relationships and practices."[72] Infrastructures, importantly, are intimately tied to "the *different and uneven conditions* that shape and characterize" them.[73] Satellites and earth-born objects in orbital and aerial space show that infrastructures are vertical as they are horizontal, and as artist and geographer Trevor Paglen argues, "The topology of orbital space is strongly influenced by geopolitical and economic policies and conventions of spacefaring nations on the earth below."[74] Without satellites and rockets, spectrum and regulation, automobiles and roads, Sirius and XM would have been nothing.

Further, the satellite's distance suggests a mysterious element, one that ushers forth myths and values that shape our understanding of the technology. But the satellite only intensifies radio's long history as an ethereal and magical form of communication. As John Durham Peters writes, "The radio signal is surely one of the strangest things we know; little wonder its ability to spirit intelligence through space elicited immediate comparisons to telepathy, séances, and angelic visitations."[75] Despite distance and expansive space, our senses are tuned to signals brought to life through consumer technologies like the radio. As radio becomes ingrained in everyday life, its ephemeral presence becomes familiar. With the satellite, however, we must listen outward and upward, beyond our field of sight, becoming conscious of what Peters calls sky media. The sky, he explains, is profoundly heterogeneous, with layers of depth, a space connected to the sea in its militarization and commercialization and their "profuse fantasies of both utopian alternatives and warfare."[76] With the satellite, our perception of ourselves and our place in this world changes considerably. Hannah Arendt said that the launch of Sputnik, "an earth-born object" sent into the universe to live with the sun, moon, and stars, was an event "second in importance to no other" and one that symbolized a "first 'step toward escape from men's imprisonment to the earth.'"[77] We experience satellites from below, as their most novel component is that they receive an uplink and deliver a downlink from way above in the sky, where we cannot touch and can only sometimes see. However, they become very real and tangible in a military context, where the power dynamic between a view from

XM's first satellite "ROCK" successfully launches on March 18, 2001.

Figure 7. Launch of XM's first satellite, Rock. *Source:* XM Satellite Radio, *2000 Annual Report*, 2001.

above can mean precision strikes from an invisible source, one of unthinkable consequences. Some satellites, perhaps in an effort to render them more relatable to the human experience, are named after birds, like Canada's first satellite launched into space (the third nation to do so following the Soviet Union and the US), Alouette 1. Translated from French to English, alouette is a skylark, a bird that flies high while it sings. XM's first four satellites (XM-1 to XM-4) were named after familiar popular music genres: Rock, Roll, Rhythm, and Blues, respectively.

The satellite is also grounded by virtue of the signals that fall on Earth. A signal that covers a geographic space on Earth is its footprint. As Parks writes, footprints are more than static maps; they are politically charged, showcasing geopolitics, trade relations, and intercultural campaigns. Footprints "are symptoms of the power of the transnational corporation in the age of globalization, in that they visualize the corporation's technological capacity to operate across nation-state boundaries, while providing little sense of the limits on this power."[78] In their growing omnipresence and power, satellites are forces for establishing, reinforcing, and normalizing values, both new and familiar, into radio, a medium thought by many at the turn of the century to be in need of improvement. As with all consumer technologies, often these qualities and values are imagined at the point of production, or when a technology is thought to be new and marketed to consumers. The satellite transforms our expectations of radio.

MUSIC RADIO FROM THE SPACE AGE TO SATELLITE RADIO

Music also communicates space. Across media and communication history, stories of musical signals routinely indicate some form of touching or communication at impossible distances. The "Earthing Project" is one example of music being used to traverse the unknown and communicate a communal human experience, or a "collective human chorus." The project, co-founded by Jill Tarter of the Search for Extraterrestrial Intelligence (SETI) and composer Felipe Pérez Santiago, involves collecting snippets of songs uploaded by people everywhere to generate a human chorus sonic signal.[79] SETI began in the late 1950s and, according to John Durham Peters, is "perhaps the most sustained examination of communication—and communication breakdown—in late twentieth-century culture," one that seeks a "true signal amid an infinity of noise."[80] Music has also been sent to the interstellar realm on two Voyager probes that were launched in 1977 and are now more than eleven billion miles from Earth. Tony Taylor, a member of the technical staff that navigated Voyager through the solar system reflected in 1989: "Now another piece of music has played out; the last note of Voyager's Neptune encounter fades into the vacuum as the spacecraft departs our solar system."[81] In these ways, music helps to familiarize outer space.

There is an idea that music communicates human nature, perhaps to those who are not human, and this is a fitting role for music given its long and intimate history with technologies of recording, reproduction, and replay. These processes crystallize ephemeral sound and enable it to be packaged, sold, and stored. Music expresses and reaches our soul, and the science and technology behind a history of recording and playback devices have indicated many ways that we make emotional connections to music. As Quincy Jones said of XM Radio, "It's a mixture of science and soul."[82] Not only is there a scientific, and potentially unknown, element to this new radio service, but something human and relatable, that of its *soul*. And its soul is music. A history of music and related audio technologies is not entirely equitable and without hierarchies, however. Technology has been linked to ideas of progress and the modern and civilized, where those who possess it are assumed to be more "developed" than those without. The concept of technological imperialism accounts for ways that one culture is juxtaposed with another through the idea that technology makes one modern, and that those without technology are premodern or uncivilized. Radio in the 1920s was connected to discourses of modernity as it became a common fixture in the home. Against the modern users of radio in America were the "racialized others," those thought to be premodern, particularly in campaigns and advertisements that espoused the necessity of owning a radio and having it reflect one's lifestyle and social standing.[83]

Although met with some skepticism about their future success (recall the Betamax comparison), there was no shortage of praise for the new satellite radio services. XM proudly claimed that its positive reviews were indicative of having "beamed the entire nation up to a new level of entertainment enjoyment, with a product of such high quality, that it has set an unbeatable standard for an entirely new medium."[84] XM introduced its letter to shareholders for the year ending 1999 by placing the service in a lineage of radio, one both simultaneously natural and revolutionary: "First there was AM. Then there was FM. And now ... XM Radio: a revolutionary new band of radio."[85] Some went so far as to relate the new radio services to the dawn of the space age. *Fortune* writer Bethany McLean said, "Sirius and XM may rattle the nation's $17-billion-a-year radio industry the way those early Soviet space successes shook America's psyche four decades ago."[86] Such revolutionary claims have attempted to validate new media for new consumers for decades. This was the case during the 1950s, when electronics and communications became affiliated with the notion of outer space as a marker of progress. "The most exciting scientific and industrial frontiers now before us," wrote *Television Digest with Electronics Reports* in 1955, "are the harnessing of the atom and the conquest of outer space—both of which will be made possible through extensive application of new advanced electronic techniques and systems."[87] Radio was thought to be an ideal medium for traversing space. The medium had accomplished great feats in this area, as signal transmission across distances is integral to its very nature. *Television Magazine* proclaimed, "Radio is, of course, indispensable to man's progress in outer space. And certainly, it will be the first human communications medium in space."[88] *Broadcasting* declared that modern radio is essential to the American way of life and that rapid and "efficient communication goes hand in hand with our national welfare and progress." Technical developments were opening "new frontiers" for transmission, including "communication with objects in outer space."[89] Following a music industry trade show in 1958, one with a notable increase in attendance due to the advent of stereo, Paul E. Murphy, outgoing president of the National Association of Music Merchants, said, "Music must take its place beside nuclear energy, space travel, communications and automation as one of the fields in which rapid strides are made." The music industry, according to Murphy, was facing a choice between belonging to "tomorrow's world, or belonging to yesterday along with the makers of carriages and buggy whips."[90] Again, the warm embrace of new technology was strategically used to define the future.

Much of the allure and appeal of the satellite as a mythic force is wrapped up in a feeling of distance and that which is beyond reach, but materials and bodies are also implicated in ideas of touching at a distance and communicating in novel ways. Space informs sound, and sound resonates in the body. In 1979 the *Globe and Mail* profiled Yale professor Willie Ruff, who said that with speakers touching bone, the music he was about to play would "go through your entire body." He was then said to have played "the music of the planets," "the sounds the planets in our solar system theoretically make as they move in their orbits," "tweeting and throbbing sounds that seemed to rotate in a dizzying way." This "celestial music project" updated the ancient idea that there is a music of the spheres by altering the notion that each planet "sang one note," instead acknowledging the elliptical paths and different speeds of the planets: "Venus hums within a range of a quarter tone, and Earth moans within a minor second."[91] Satellites are vehicles for channeling the ethereal sounds and music of space. The first sounds humans heard from space "were the 'beeps' of the orbiting Sputnik satellite picked up by amateur radio enthusiasts," and thus electronic sounds, drones, beeps, and blips came to signify space and its vastness.[92] In 1960 Toshiba Records "brought the satellites down to earth" on a recording of electronic beep signals. This was a 10-inch LP that included radio signals from American and Russian Earth satellites that had been in orbit since 1957.[93] Blips, beeps, and instruments like the theremin color our perception of space, as does music. Ethnomusicologist Timothy D. Taylor characterizes space-age pop music as "jazz-influenced popular music of the 1950s and early 1960s that thematized the exotic, whether terrestrial or in space, and was intended to be played on hi-fis."[94] This music reflected a broader postwar obsession with the future, which brought new technologies into the home as part of an effort to convince people of the "peaceful use of atomic power."[95] A series of album reviews from the late 1950s and early 1960s charts this fascination with the music of outer space. A review for "Stereo-Only Albums" in *Billboard* in 1958 described the album *Music for Heavenly Bodies* as music that, through the novel use of the electro-theremin, was "meant to convey the feeling of flight into space."[96] A 1962 review of the album *Music from Outer Space* explained, "There have been many 'outer space' musical records, but this one is not only musically attractive but also contains a gimmicky collection of space sounds that add a nice touch to the recording."[97] Some of these albums put the sounds of satellites front and center. Satellites, one of "the most talked about events of the day" in May 1958, had songwriters conjuring the novel technology in songs like "Satellite Be-Bop" and "The Sputnik Story," as well as other "strange songs" that "concern space people and moon people who land here and scare us Earthites."[98]

Throughout the space age, those complex technologies of the space race and the research and development that went into military weapons following the creation of the atomic bomb and into the Cold War, often shared much with new consumer technologies. A line of Philco transistor radios from 1957–1958 was advertised in stores, with counter displays including a panel that connected the Philco transistor line to the Vanguard satellite, with copy that explained that the transistors were used in both the satellite and the portable radio.[99] Novel technologies underwent processes of familiarization and humanization, a trajectory that Taylor explains using the concept of "commodity scientism," which is a belief in the "ineffable qualities of science and technology," an ideology that arose as a result of the increasing prevalence of technology, often difficult to comprehend, in American lives.[100] One humorous example of this with respect to satellites was vending machines remodeled to resemble the American communication satellite in Belgium: "Globe-shaped vending equipment is altered and painted to resemble Telestar [sic], and miniature Telestar [sic] charms are inserted with the ball gum and nuts." Experiments were conducted to have the machines emit "outer space sounds resembling the transmitter signals of space satellites."[101] Radio helped make sense of space age technologies for earthlings. A research project out of the Stanford radio propagation laboratory in the 1950s was working to make "outer space regions, far beyond the limits of the known ionosphere, accessible to study by radio for the first time in the history of science."[102] Popular records of educational fare figured into this process as well. Folkways Records' *Voices of the Satellites!*, a documentary album for "science buffs, students, educators," featured radio "voices" of Russian and American satellites, complete with "extensive" notes.[103] *A Child's Introduction to Outer Space* by the Satellite Singers offered a "primer in the fields of astronomy and astronautics" set to a "perky Latin rhythm" and narration about how satellites were launched.[104] Across the technologies of radio, satellites, and recorded music, outer space has become mediated and commodified for earthbound consumption.

There is evidently a long history of hearing outer space in music, and this was certainly true at the turn of the millennium when Sirius and XM began, offering a sonic accompaniment to this particular nexus of music, technology, and industry. There are numerous examples of satellites and themes of outer space being manifest in music, thus shaping value judgments about music and, by extension, the processes by which music circulates. The satellite has figured into numerous popular songs, whether through lyrics or through the use of experimental sounds and instruments. Elliott Smith, Rise Against, Spoon, Kevin Gates, Thundercat, Sara Hartman, Petra Marklund, and Nine Inch Nails all have songs titled "Satellite." In 1962 The Tornados hit number one on the US *Billboard* Top 100 with "Telstar," an instrumental track that featured an electronic keyboard and was named after the communications satellite that was launched the same year. Lou Reed's 1972 single from *Transformer*, "Satellite of Love," may be one of the most well-known songs to feature the satellite. It ends with the words "satellite of love" repeating over and over again as it fades, with Ziggy Stardust-era David Bowie providing background vocals. Low's *Hey What* (2021) begins with "White Horses," a song that opens with what sounds like a massive machine awakening with tension and the grinding of gears, making an effort to launch. A staccato electronic signal and angelic vocal harmonies then descend on the track. As the song ends, a pulse becomes higher in pitch, a repetition like radar but with tempo increasing, sounding like systems are failing. As with infrastructure, we become aware of the machine in its failure. Across the album, the voices of Alan Sparhawk and Mimi Parker are clear and often echo, simultaneously filling the space of headphones while spiraling outward into the unknown. Behind them, repetitive blips, tones, agitated feedback and explosive drones characterize the ten songs, which bleed into one another. It sounds like a satellite mediating outer space for earthbound ears.

Scientifically, there is no sound in space, but by its very nature, music fills space, an intriguing paradox to be sure. Musicologist Ken McLeod has indicated that a fascination with themes of space, aliens, and the future involves a fascination with the unknown and unidentified.[105] This is perhaps all the more reason why space, planets, the stars and moons, UFOs, and alien lifeforms are frequently heard and seen in popular music history. Ideas about innovation have connected space and satellites to genres and styles of music that are defined and understood as cutting edge. One profile on the new jazz age of the 1950s explained that the ears of younger listeners between ages seventeen and twenty-seven are naturally tuned to the new jazz sound, just like Earth satellites and rockets are to outer space, whereas the older generation's ears are simply hearing dissonant sounds.[106] The launch of Sputnik brought the phrase "space age" into the poetry and music of Sun Ra, a legendary musician for whom space "was the place where better living becomes possible through music"; it "made space real" and "provoked the *invention* of outer space as an international if contentiously collaborative creative enterprise."[107] Artists like George Clinton and Lee Perry created otherworldly environments, "futuristic environs that subtly signify on the marginalization of black culture," connecting diasporic African history to "a notion of extraterrestriality."[108]

Ethnomusicologist Michael Veal aligns the reverberation and delay of dub music to the vastness of outer space and argues that the "otherworldly strain in dub music partially evokes an idealized, precolonial African utopia."[109] Because space is without sound, there is room to imagine what it could sound like, what it can signify.

The advent of space rock would also, rather obviously, draw on themes of outer space and space exploration to chart ideas of sonic expansion and challenges to conventions in rock music. As music writer Jon Savage explained for *MOJO* in 1995, "Space has provided a place to unfetter the imagination, to push the limits, to do weird shit." He added that the theme "occurs most heavily when there is a technical/perceptual shift," citing examples of Elvis and the transcendence of acid house.[110] Rock's growing significance in culture throughout the 1960s and 1970s mirrors the intensification of the space program in America, particularly with the first live television broadcast by Telstar in 1962 and the Apollo moon landing of 1969. Pink Floyd's *Dark Side of the Moon* evokes this realm by its very title but also in the prominence of synthesizers, studio production, and sonic experimentation.[111] Rock's spacey subgenres of space rock and progressive rock, with atmospheric sounds, guitar effects, and "banks of keyboards, myriad of knobs and dials," were, according to McLeod, "roughly analogous to the technology being developed and exploited in the real space programme."[112] The UK space rock band Hawkwind's "Silver Machine" had lyrics inspired by Alfred Jarry's essay "How to Construct a Time Machine," and its album, *In Search of Space* (1971), was accompanied by a twenty-four-page "Hawkwind Log" that refers to a spacecraft occupied by artists searching for extraterrestrial intelligence. Another artist who has relied on space and alien imagery to advance ideas about innovation, mastery, and hip hop authenticity is American rapper Lil Wayne. His meteoric rise in popular music has undoubtedly been marked by the success of 2008's *Tha Carter III*, which followed a series of critically acclaimed studio albums (*Tha Carter* [2004], *Tha Carter II* [2005], *Like Father, Like Son* [2006], and a number of mixtapes) that landed during the years in which Sirius and XM attempted to reach a sustainable subscriber threshold. A 2009 *Rolling Stone* profile of Wayne noted that *Tha Carter III* was the best-selling album of 2008 and that he made guest appearances on "a staggering 110 tracks by other artists."[113] *Tha Carter III* and Wayne were characterized by a theme of the extraterrestrial. Lil Wayne referred to himself as a Martian and an alien, an otherworldly being who was above and beyond the skill level of other rappers.

On the one hand, the recurring use of space and alien imagery and themes has encouraged the proliferation and empowerment of marginalized identities. As one example, McLeod argues that the iconic persona of Ziggy Stardust, David Bowie's alter ego for the 1972 album *The Rise and Fall of Ziggy Stardust and the Spiders from Mars*, was meant to illuminate the artificiality and heterosexuality of rock music.[114] On the other hand, the trajectory of space-themed records, from space age pop music to space rock and beyond, has often envisioned men as the primary listeners, particularly through a perceived idea of masculinity as being constructed and expressed through a technical mastery of new technologies and the spaces they provided access to, real or imaginary. One review for a spoken word album titled *Project Moon* from 1957 explained that the "disk [was] gimmicked with sounds of a rocket and various electronic equipment" and that it likely only had "varied appeal with greater interest, perhaps, to hi-fi bugs."[115] These hi-fi "bugs" were predominantly men in the postwar era, and the mastery of hi-fi players and domestic space devoted to the listening of music on the hi-fi player became a way in which masculinity was constructed in the primarily feminine space of the home.[116] The hi-fi player was a source of exceptional sound that could fill domestic space and transport the listener elsewhere. The satellite continues this narrative. One preview of the developing Satellite CD Radio system made this link quite explicit, quoting from CD Radio president and CFO Peter Dolan, who said that the system would "enable a broadcaster to transmit a true digital signal to the listener. . . . [I]t would be comparable to listening to a CD player."[117] Before the hi-fi enthusiasm of the 1950s, early radio broadcasting involved early adopters who enjoyed experimenting with signal transmission and radio set building, efforts at traversing distance through wireless. Often these radio enthusiasts were young boys and men, before radio became domesticated and licensed. Radio historian Susan J. Douglas has written about the "rise of the boy inventor-hero as a popular culture archetype" in radio and the control or mastery of technology in the early 1900s.[118] A fascination with radio's ability to shrink great distances carried forward into the space age and beyond, with ongoing interest in sending voice signals to space.

On the fortieth anniversary of the Apollo 11 mission, for example, amateur radio "buffs or 'hams'" held a global "bounce-fest" using giant parabolic antenna radio telescopes to send a signal to the moon and back.[119] Expressions of exemplary sound quality and the theme of exploring the boundaries of new technology would also fetishize and exoticize women, proclaiming these spaces as masculine domains. The aforementioned review of *Music for Heavenly Bodies* concludes by stating that "Omega's fancy packaging features a cover with a lovely nude female floating in outer space. Display is sure to aid sales."[120] This masculinization of technology was characteristic of the early programming of Sirius and XM as well.

In their conquest to occupy the sonic space of the vehicle and convince listeners about a superior radio technology, Sirius and XM were prone to imagine an ideal consumer for the service who was a young adult, employed, and with enough disposable income for a subscription, one who also was typically male. Early press releases about the new lifestyle channels offered by the services revealed men's interests, like the Maxim channel, which aimed to "reflect the sex-packed articles and down-and-dirty humor of the publication."[121] XM also had a Playboy channel that came with an additional cost to subscribers, and Pamela Anderson was a Sirius host and early spokesperson for the brand.[122] In one interview with Sirius dance and hip hop music programmer Geronimo (Jonathan Broth), he says, "We just hired Pam Anderson to host her own show, and, quite honestly, I'm not mad at having her around the office."[123] The majority of the celebrities and musicians who would be behind specialty channels on Sirius and XM were also men. In 2006 it was noted that XM was looking to develop more programming aimed at women, who had been slower to adapt to satellite radio than men.[124]

The values of fidelity, quality, innovation, and experimentation are tied to larger cultural contexts and themes with respect to age, class, gender, and so forth. Other supposed values of the satellite involve its capacity to collapse both space and time. Combined, these characteristics shape ideas about the production and circulation of music. This ushers in a consideration of geopolitics and the power dynamics between and across nation-states, especially given the borders of the satellite's footprint.

With the space age came idealistic aspirations for a much more international music industry, with cross-border connections between various industries and artists, and the technology of the satellite was routinely central to these stories. In 1962 *Billboard* profiled the "future in record merchandising," one that was said to lie in mergers between distributors and that saw the record industry headed "toward a 'one-world' concept," driven, in part, by communication media like the Telstar satellite. According to Mercury Record Corporation president Irving B. Green, this meant that "foreign artists are becoming increasingly important to the entire industry."[125] Years later, MIDEM commissaire general Bernard Chevry anticipated that the 1970s would see "more and more people working in the industry ... working together without regard for national barriers: because the world is now covered by radio and television and because we live in an age of regular communication by satellite, it would be absurd to preserve the old spirit of regionalism."[126] Some examples of these satellite-facilitated international broadcasting practices and events are a radio DJ exchange between Australian and American RKO Radio stations in 1976, in which the DJs would travel to the other nation and broadcast programs back home via satellite,[127] and a Jethro Tull concert at Madison Square Garden in October 1978, which was "beamed via satellite" to Europe and the UK to reach nine hundred million people, an event that led program director of Toronto's CHUM-FM Warren Cosford to proclaim that "the future is limitless."[128] In 1981 it was said that *satellites* was the big buzzword at radio programming conferences, and some believed the technology would "pave the way for great mass appeal 'national' stations to emerge ... sort of the fulfillment of McLuhan's 'global village' philosophical prognostication." One of the imagined uses of satellites would be to have the "'best' disk jockeys and air personalities staffing a relative handful of 'stations in the sky.'"[129] Satellites and their international reach brought forth ideas about democratizing the gaps between rural and urban listeners, between those who lived in cities and those who were more isolated. One example explained that a listener in Alaska would receive the same quality of signal as one in Manhattan.[130] A print ad for Sirius from 1999 read, "Greybull, Wyoming. One Traffic Light. 100 Radio Stations." The ad copy continued to say that there was "no escaping" satellite radio; it would "reach nearly every driver on every highway, byway, main street and back road."[131] These utopian aspirations followed decades of loading the satellite with such lofty goals.

Figure 8. Sirius ad showing Greybull, Wyoming. *Source: Radio & Records*, November 26, 1999.

Satellites, however, are very much bound to borders and boundaries. There is a high degree of control when it comes to monitoring the boundaries of the satellites' transmission coverage areas, or footprints. In Lisa Parks's groundbreaking study of the culture of the satellite, she explains that the satellite can produce a photo of the "whole earth," but satellite television practices have divided the world in "ways that support the cultural and economic hegemony of the (post)industrial West."[132] In 2006 XM described how its satellite footprints "encompass the 48 contiguous states, nearby coastal waters and the densely populated regions of Canada" that are tucked close to the American border.[133] XM Canada, XM's licensee, had acquired a broadcast license at this time, which explains the signal crossing the border to the north, but transmission spillage was tailored to minimize spillage south into Mexico. Coastal waters coverage enabled the services to target cruise and cargo ships and leisure boats. The satellite's footprint, then, is shaped and controlled by policies and practices tied to power dynamics rooted in geopolitics and political economy. Processes of distribution involve standards and formats that direct and deliver content; the resources, technologies, and labor required to distribute audiovisual signals on various scales from the local to global; and the policies crafted and the parties involved in policymaking for shaping how infrastructure is developed and used.[134] The US, for instance, has maintained a strong presence and level of control in the realm of satellites and orbits, and through policies and strategies, a nation is able to protect the asymmetries of economic and political relations that preserve its power.[135] In 1973 the ITU declared the GEO a "limited natural resource," which demarcated the orbit, legally, from space in general.[136] As with other natural resources like terrestrial radio spectrum, there is limited availability of frequency bands and orbit positions with high demand.[137] Nation-states have been tethered to these terrestrial geographies in ways that highlight the uneven power between them. In 1977 it was noted that the US had been "singularly successful" in achieving its aims at the international conferences of the ITU and had over 95 percent of its proposals adopted.[138]

In addition to geographic expansion and coverage through transmission and communication, the satellite also brings forth so-called revolutionary ideas about collapsing time. The 1967 international television broadcast of *Our World* advanced ideas of live satellite television and its production of a "global now," although it was scripted, planned, and rehearsed.[139] One advertisement for Visions Mobile asked in the wake of the American Music Awards in Los Angeles: "How did Boy George get from Mornington Crescent to Sunset Strip in a tenth of a second?" The answer was that a satellite could move the broadcast over fifty thousand miles in one-tenth of a second.[140] In 1989 Satellite Media Services made use of the Intelsat satellite to distribute mono or stereo audio and data from London to equipped radio stations anywhere in northern Europe. The satellite transmission was described as "a fast and direct method of distribution, offering virtually instant delivery and confirmation."[141] Collapsing time by satellite was viewed by some as such a monumental achievement, and one that was cutting costs for radio production and distribution of remote shows, that IDB Communications Group asked in 1988, "Why are we pressing all this vinyl when [a remote show] can be sent quickly with better quality via satellite?"[142] The *instantaneous* nature of communication by satellite technology mirrors the sort of value assigned to a notion of liveness. This is a point made by Philip Auslander in his critical assessment of liveness, in which he connects liveness to an idea of authenticity that circulates within rock culture and gives it value to its participants.[143] As of the 1920s, "live" meant "simultaneous broadcasting";[144] over the course of a hundred years, our need for faster and more instantaneous communication has continued as we have become accustomed to a particular speed at which we expect the transmission of data and information to occur. Time is central to music as well as the rhythms and circuits of our daily lives, which are often accompanied by music. An early rock 'n' roll song like "Rock around the Clock" by Bill Haley and His Comets makes reference to both space and time in the objects of the comet and the clock. Just as we feel music in space, we do so in time, and a technology that can expand the range of music delivery by radio, as well as the speed by which audio signals are received by listeners, is crucial to acknowledge in the history of popular music.

ON ORBITS

Collectively, the values and qualities embedded in our understanding of the satellite and its operation encourage a wider understanding of radio and music in the twenty-first century, time marked by digital and streaming music and listening technologies becoming dominant in everyday life. One way to conceptualize our music culture is in the notion and nature of orbits themselves: in rotation and cycles, and in a fascination with that which is distant enough to feel special and unique but also present and tangible at the ground level. The "world of orbits" is found at the margins of Earth's atmosphere and "on the threshold of the vast realms of space." It is where we encounter the "crucial but neglected manufactured environment of satellites and space junk."[145]

Records are synonymous with rotation, as is the orbit. A turntable spins, as does a DJ. A CD whirs in rapid rotation while a laser reads its data. The gears of a cassette turn magnetic tape. Satellites are in orbit and in sync with the earth, covering a portion of the world with their beams. This connection between orbits and recorded sound was made explicit with a "Project Moon" EP package from the aptly named Orbit Records in 1957, one featuring a "barrage of unusual sound effects" and "recorded in full spectrasonic sound."[146] Orbits are also about time. As Paglen writes, spacecraft in the geostationary orbit are "not only part of the world's communications backbone, but will invariably be some of the humans' longest-lasting artifacts." The orbital belt is so far away that these spacecraft do not experience orbital decay and could potentially remain for billions of years into the future.[147] Orbits, then, encourage questions about how certain recordings persist through time against all odds. What forces are at work to keep certain songs, albums, or artists in the cultural imaginary, and what forces relegate others to be lost, like space junk?

Satellites in the GEO rotate in sync with Earth, suggesting cycles that we can feel, relate to, and harmonize with. In this way, the GEO is an apt metaphor for thinking through our relationship to music across time and space. It helps to ground us in some way within an expansive world and universe but also makes sense of the ways that music enters and leaves our lives in ways both mysterious and predictable. Through media and technology, we can access the orbit, and our lives on Earth are shaped from above. The role of orbits in conceptualizing the music programming of Sirius and XM, and SiriusXM, is taken up in further detail in chapter 3, but for now, I simply wish to introduce the idea with respect to the satellite, its place in outer space, and the values it ascribes to understanding music radio more broadly.

To harness the power of orbits requires rockets, satellites, and millions of dollars in research and development, as well as time. For Sirius and XM, the satellite has been central to their longevity due to the perceived value it lends the service and the music it programs. In a *New York Magazine* profile of satellite radio from 2000, writer Michael Wolff argues that "technology *guys*" (emphasis added) are drawn to media because they know that pipes and optics and cells are the "important stuff." He adds, "If you can't talk pipes or satellites, then you have no business in show business. The smart media-business plan of the future will be about reaching an audience by building a new system, an amazingly complex and costly system . . . that will be better than the existing system it bypasses."[148] Sirius and XM built the costly system, and they acquired licenses to do so. Whether the system is truly new or the adaptation and reworking of already existing systems is a point to debate, although I argue that their success in launching was very much the result of establishing key partnerships between existing businesses and industries and finding the investment to build, launch, and control their own satellites and network of transmitters. To deliver on these massive investments, both Sirius and XM would soon face the urgent need to compete for subscribers. This is the topic of the next chapter.

Writing on cultural technologies, Berland refers to Theodor Adorno's essay on mediation and classical music, in which he discusses the development of composers facing an anonymous marketplace and needing to sense a demand in the market, an "embrace between creator and market." Berland argues that "cultural technologies can be understood as versions of this same productive pursuit. As this process is entrenched within the culture industries, the search for listeners leads to increasingly rationalized methods for identifying their situations and subjectivities, and for formulating specific modes of address to please them."[149] Berland ties the need to find new viewers and listeners to Harold Innis's idea of space-biased media, which favor "dissemination across space over continuity in time."[150] The satellite persists through time, as an object in space that lives on past its useful life. And satellite radio mediates time through an expansive list of channels that feature music from the past, such as channels dedicated to the 1940s to the present. The satellite's footprint, however, in its spatial advantage to the transmission range of a broadcast tower, is a crucial aspect of the technology given the listeners, or subscribers, who fall within it. Without subscribers, Sirius and XM would be doomed. Both companies knew this, and they would compete heavily for them throughout their formative years.

2 Targeting Subscribers in Satellite Radio's Formative Years

Included in the glovebox of my 2013 Chevrolet Cruze is a safety-yellow envelope that includes a quick-start guide with instructions on how to use XM Radio during a ninety-day complimentary trial. Six individual caricatures of imagined listeners decorate its front cover: a relatively nondescript middle-aged man, a composer, a sports fan, a NASCAR fan, a hip hop artist, and a heavy metal guitarist. Inside the booklet are instructions for operating the radio service, but more importantly, so too are convincing accolades about the act of listening to XM Radio: "XM is a truly unique listening experience"; "driving will never be the same"; "Just coast to coast coverage of digital sound, all the time. It's like having your own personal digital music player, but without spending hours (or dollars) downloading." The booklet's NASCAR fan represents a marketing and programming partnership between the motor sport company and XM Radio. But why partner with NASCAR to be the home of radio broadcasts of races and related coverage? The answer lies in the need to acquire subscribers, regular and reliable sources of income who must be convinced of the value and worth of their ongoing subscription dollars. The growth and sustainability of satellite radio was dependent on subscribers, who would be courted, in part, by a balance between niche and mainstream channels. In

2003 subscription revenue comprised over 85 percent of XM's total revenues.[1] And in 2011 the merged SiriusXM generated 98 percent of revenue from subscriptions.[2] High subscription revenues distinguish satellite radio from its competitors in broadcast radio, which rely on advertising dollars. A key question asked by the trade press was whether radio listeners could be made subscribers.[3] Given the comfortable relationship between popular music and CDs throughout the 1990s, many wondered whether the impetus for people to own things was "too important for subscriptions to become dominant."[4] Sirius and XM would need to convert radio listeners to radio subscribers and convince them that radio was worth paying for.

In their respective paths of development, Sirius and XM established their services in spaces that, in opposition to the wide accessibility of FM and AM radio, were exclusive and restricted to middle- to upper class listeners. The *ideal* subscriber-listener is defined by class and cultural hierarchies, and we get a sense of this in Sirius's and XM's focus on the space of the automobile, the airplane, luxury hotels, and coffee shops. In other words, a particular class of subscriber-listener was *targeted* by the companies. The automobile, an "everyday object where human beings regularly encounter new technologies in their everyday lives and learn to 'inhabit technology,'" was an especially crucial technology for convincing listeners to become subscribers, namely through factory-installed receivers and free trial periods.[5] The early years of Sirius and XM tell us much about the numerous subscriber-based media services that would follow in the wake of satellite radio. Like its now-competitor Spotify, SiriusXM has worked to "maintain the image of a company that is always expanding, looking forward" in its pursuit of media concentration and monopoly position.[6] *Broadcasting Magazine* claimed in 1957 that the age of "satellites, spaceships and guided missiles will require constantly expanding electronics production."[7] The satellite radio services that developed out of this constantly expanding production would also seek out a constantly expanding subscriber base.

"PREMIUM" LISTENERS FOR "PREMIUM" RADIO

Radio has a long history of being understood as a free and accessible medium. Commercial broadcasters are funded primarily through advertising, with public and community stations receiving funds from the government via tax dollars or from listener donations. The initial monthly subscription cost for Sirius and XM was just under $10 a month—$9.95 to be exact (today, plans range from the app-only cost of $9.99 a month to the Platinum plan of $23.99 a month). "Premium" radio would require "premium" listeners, those with the disposable income, and as advertisements would suggest, musical tastes and preferences that were beyond those of the average commercial radio listener.

One factor that helped to position the new services against "old" radio was that industry consolidation and capitalization had constricted music playlists with more standardization across stations and markets. From 1990 to 2000, a total of 900 separate companies that owned radio stations fell to 720, and the trend would only continue. Clear Channel Communications, one company implicated in the intensifying consolidation at this time, owned more than one thousand stations in 169 markets across the US in 2000. Consolidation was said to be a "plague on programming," resulting in a "more homogenous radio dial."[8] This meant that the landscape was ready for the new satellite radio services to promote their features and that there was a chance listeners could be convinced to purchase a special receiver that cost around $120 and also to pay the $10 monthly subscription fee. Music channels would be free from advertisements, although some music channels on XM initially carried ads.

At the turn of the century, much radio-centric discourse within the industry was focused on change. In 1999 some reports claimed that radio was in a prime position to take advantage of "the new media age," having "the greatest potential the industry has seen in its 80-year history" due to its strong financial stance, the potential of the internet, and the development of digital radio.[9] Whether satellite or terrestrial radio, digital media was the future, and it was time for the radio industry to catch up. With the massive investments in time, research, and money that Sirius and XM had put into launching satellites and establishing receivers, industry partnerships, offices, and personnel, the two companies were in a sound position to capitalize on this theme of change. CD Radio's executive VP of content, Joe Capobianco, was confident that the company would have forty-two million listeners within its fifth year, "making it the fastest-penetrating electronic medium in history."[10] CD Radio targeted

TARGETING SUBSCRIBERS IN SATELLITE RADIO 63

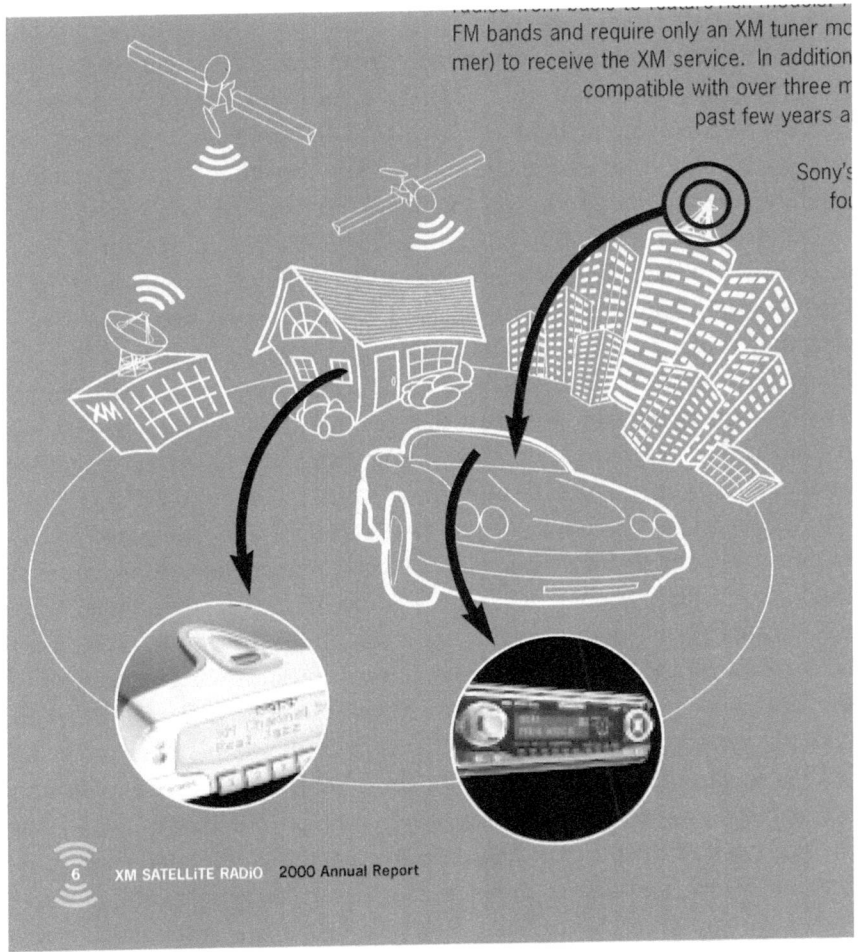

Figure 9. Graphic depicting XM satellite radio's reach. *Source:* XM Satellite Radio, *2000 Annual Report*, 2001.

broadcasters in advertising copy and was "going after anyone who is disappointed with traditional radio."[11] XM explained that it aimed to focus on eighteen- to thirty-four-year olds at first, because of their status as early adopters of new technology. By 2006 Sirius and XM had a combined subscriber base of just over eleven million, *just a few shy* of the forty-two million Sirius had eagerly anticipated in 1999.[12]

Across trade press reporting on the early years of Sirius and XM, themes of exceptionality, newness, and distinction characterized the services compared to commercial radio. At a trade convention for the 30th Country Radio Seminar held in March 1999 in Nashville, Lee Abrams of XM called FM "tired and cliché-ridden" and said that the forthcoming satellite service wanted *fans*, not listeners. Abrams added, "We're going to be national and proud of it and give listeners the impression we get records first. When there is a new release, we're going to celebrate it, not just play it."[13] One Sirius ad campaign from early 2003 derailed the notion that satellite and terrestrial radio could have some sort of "peaceful coexistence." The print, outdoor, and television campaign was framed as a "musical manifesto," with text that read: "Payola_OFF ... how many palms have been greased?"; and, "Commercials_OFF ... music shouldn't be brought to you by a double espresso in a can." A "radio_OFF" page took aim at the limited broadcast range of terrestrial radio, compared to the nationwide coverage of Sirius, and said that "censorship is more profane than a few profanities."[14] A sense of dissatisfaction with commercial radio was not unique to the mid- to late 1990s and into the 2000s. One report by the Canadian Radio-television and Telecommunications Commission in Canada from 1975 advocated for using the FM band to diversify the country's broadcasting system and said that numerous complaints were being leveled at commercial radio, such as claims that it programmed trivial and uninteresting content that was limited in scope.[15] The BBC's Radio 1 has defined itself against the music of commercial radio, liberated from commercial criteria and able to program music from more specialist genres.[16] Listener-supported stations like Seattle's KEXP are strategically positioned against commercial radio through enhanced offerings like video streams of performances.[17] Freeform FM stations in the US have also been distinguished from commercial radio, as well as musically dynamic campus and college stations across North America.[18] These critiques are part of a longer trajectory of ad-supported, commercial cultural products being thought to have limited creative and artistic value, and they reflect a longer history of people advocating for more experimental and innovative uses of radio.

The business strategy of relying on subscriber payments to fund *premium* radio is a notable shift in terms of conceptualizing what makes radio

radio. Namely, it alters the nature of radio's accessibility, with subscription money becoming a barrier. In Raymond Williams's cultural study of television, he says that radio and television broadcasting "was obscured by its definition as 'mass communication': an abstraction to its most general characteristic, that it went to many people, 'the masses', which obscured the fact that the means chosen was the offer of individual sets, a method much better described by the earlier word 'broadcasting.'"[19] Williams challenges the articulation of a mass and the technology of broadcasting, given the private listening contexts surrounding television and radio. In other words, radio has always targeted listeners in some way, even when "free" or supported by advertising. While the basic formation of a centralized hub of production that transmits programming to disparate listeners, whether as sets in a home or as individuals on the move, is more or less unchanged in the satellite radio universe, an emphasis on the subscriber-listener points to transformations in audiences and listening communities as well as an increasing emphasis on the *targeting* of individuals, from sets and receivers to unique taste preferences and niche demographics.

Radio broadcasting is incredibly influential on the formation of imagined communities on local and national levels. Broadcast historian Michele Hilmes, using Benedict Anderson's concept of the modern imagined community of nationhood, writes that this concept is absolutely alive in "listeners' tuning in by the tens of thousands to one specific program airing at a specific time," creating a "shared simultaneity of experience."[20] Hilmes adds that in radio's reach of a vast, invisible audience, the "choices made by early stations, networks, sponsors, and agencies as they invented themselves and the 'business' of radio reflect the tensions of a diverse and divided society." The reflection of the social order, then, can change depending on funding models, broadcast range, corporate cultures, and so forth. Syndicated, national, network programming of the 1930s, funded by advertisers, anticipated and imagined a different audience than a local campus-community radio station in Sackville, Canada, broadcasting to a few thousand people in the 2010s.[21] And certainly a community of listeners choosing to pay a monthly fee to have access to premium radio is distinct in its own way, particularly when business strategy and corporate culture are shaped by the number of subscribers as opposed to the drive to appease advertisers. With satellite radio, to subscribe, drive, and be mobile is to be included in the listening community. Sirius and XM's coverage is not so much national as a sort of boutique nationalism made up of subscribers, from which the companies can consistently and predictably extract profits.

The pursuit of premium listeners, or subscribers, calls attention to the function of the technology of the satellite. The photo imaging capabilities and military and surveillance uses of satellites apply to the ways that the satellite companies target listeners or isolate profiles of listeners for the sake of crafting programming. Lisa Parks explains that the term *remote sensing* refers "to a practice of orbital viewing that simultaneously extends and exposes particular regimes of vision and knowledge."[22] Remote sensing was carried out extensively during the 1960s by the US for monitoring the Soviet Union, Eastern Europe, and Asia.[23] After the September 11 attacks in the United States, and also around the time of the launching of Sirius and XM, the global war on terror meant that media technologies, including satellite images, were tuned to issues of global security. In the years following, an array of public and private agencies collected more information "than anyone could possibly comprehend."[24] The satellite became a key component in the pursuit of information and in endless surveillance at the expense of personal privacy.[25] Drawing on Martin Heidegger's notion of the world becoming a picture, literature scholar and cultural critic Rey Chow suggests that in "the age of bombing," the world has become conceived as a target.[26] Across the Gulf Wars came the "ubiquitous virtualization of everyday life," with those above having "privilege of access to the virtual world" and those below tied to "the random disasters falling from the heavens."[27] American studies scholar Caren Kaplan describes how "air power" affords an advantage to military powers, who control aerial vision technologies and have the ability to surveil others.[28] The wider context of the Gulf Wars and war on terror is inescapable when thinking of the role and purpose of satellites in the 2000s. Even though the focus here is on satellites and radio, the satellite is a technology that targets. As subscribers become essential to the function and operation of satellite radio services, the target implies a relationship between what the satellite can accomplish and who the satellite radio listener is or will be, a zooming in on the ideal consumer for the subscription radio product. In Chow's formation, the world conceived as a target is an object to be destroyed. However, for Sirius and XM, the vast numbers of music and radio listeners across the US were targets to be sought out and convinced of the virtues of this new type of radio, were sources of revenue to be extracted.

MUSIC SUBSCRIPTION BEFORE STREAMING

A shift to subscription models for music consumption is often thought to be most obviously bound to the dominance of music streaming platforms in the early 2010s in North America. A turn to services like Spotify and Apple Music meant that listeners were less connected to material copies of music. Instead they paid a monthly subscription fee to access extensive music catalogs stored on servers and clouds through a wireless internet connection and a mobile device. Sociologist David Arditi remarks on this move to subscription services as evidence of the unending consumption of music, whereby consumption becomes consistent and constant and subscribers forgo music ownership.[29] Digital music subscriptions, he adds, resemble cable television subscriptions in the sense that when the subscription ends, one is without access to content and without any material cultural products. Sirius and XM would take on the cable television subscription model, but importantly, they set the stage for the broader turn to subscriptions within the world of popular music. Subscription radio is intriguing given that radio has always been separate from album consumption, aside from the fact that it promotes the sale of albums and that albums have been used for on-air programming. Within the satellite radio model of the subscriber-listener, the decision to subscribe is arguably less about unending consumption, although the nature of the subscription service facilitates the same economic logic, and more about the desire to target one's attention via taste and musical preferences in an era of seemingly endless choice. Information society brought with it an "over-abundance of information of all kind[s]."[30] Through subscribing, a listener finds solace and stable footing within the "unimaginably unmanageable flow of mediated information" in "an era of information glut,"[31] which has become available to anyone with access to the internet, digital music files, online radio, and streaming music platforms. This feeling of being overwhelmed with choice is not limited to information overload. As Amanda Montell writes in *Cultish*, there is a burden of independent choices to make about what we think and who we are. Thus, the following of cult brands like Lululemon, Starbucks, or even a cult studio fitness company like Soul-Cycle, which has a SiriusXM channel, or Peloton, which uses a subscription model, can help one to cut the overwhelming number of choices one is faced with down to a manageable few.[32] The radio channel and its host will do the work for us; we just need to decide where to turn the dial.

One way that we can think of value being ascribed to music in a nonmaterial context is through subscriptions, particularly those with a high cost to access exclusive content. People tend to accumulate subscriptions, or at the very least, subscription logins and passwords. To return to the metaphor of orbits, a collection of subscription services encircles us, enabling the consumption of news, film, television, music, and so forth. The more access one has, the more culturally current or relevant one is able to be. We might think of the question, "What's in your orbit?" as reflecting the capacity to be well-versed in media and cultural objects, a disposition that is increasingly shaped by access to subscription services. Most media consumers maintain some balance between their own subscriptions to film, television, music, or journalism companies, and logins or subscriptions shared among family members or friends. In 2020 it was reported that nearly eleven million people in the US accessed a major streaming service through shared login information without paying for a subscription themselves, and 44 percent of all millennials did.[33] Netflix, Hulu, Amazon Prime, Spotify, Apple Music, YouTube Music, SiriusXM, Peacock, Crave, and Disney+ are just some of the many streaming media services North American consumers subscribe to today. To manage the demands of work, parenting, and busy social lives, people may also subscribe to services that promise to make everyday tasks simpler, like at-home meal prep delivery services.

But for radio, and even for music listening more broadly, subscription was not a common mode for listening in the late 1990s and early 2000s. In 2002 a short profile of new music subscription services, which included XM and Sirius but also internet services like MusicNet, Pressplay, and MusicMatch, indicated industry skepticism about whether the business model and consumer proposition of subscription could work for the music industries.³⁴ Would people willingly pay for radio while at the same time having to pay monthly phone, internet, and cable bills? Is radio in the car all that necessary when drivers have access to CD and cassette players? One editorial asked, "How Many More Monthly Fees Can Consumers Stand?" The question posed earlier, "What's in your orbit?," is not without its limits. Debate centered on whether downloads or subscription services were the way of the future. Those in the subscriber camp were keen to treat music like a utility, with users having access to all the music they want, thanks to a monthly fee. But this meant that the subscription services would be competing with "a battalion of other media and entertainment subscriptions that in recent years have crowded their way onto credit card statements."³⁵ Mark Mooradian, the senior director of strategic planning and business development for the internet music service MusicNet, has said, "Treating music outside the realm of ownership is definitely a new concept for consumers and something that eventually requires some behavioral change."³⁶ In reality, the concept is much older. Jacques Attali's *Noise: The Political Economy of Music* highlights the fact that owning music-as-commodities is truly a recent and temporary moment in music's much longer history.³⁷ But Mooradian would be right in the sense that consumers, at that moment in time, were accustomed to purchasing, owning, and collecting material copies of recorded music.

Other music subscription efforts were attempted before the 2000s, though with limited consumer uptake. In January 1970 Israel Diamond, systems coordinator for Peer-Southern, detailed the context of data processing that surrounded music listening at a time when music was "more listened to for more hours of the day than any other sound since history was recorded." Diamond explained how mechanical reproduction and the storage of sound corresponded to the development of copyright law and data processing that enabled sampling systems to evaluate the number of times a work was credited with performance. Diamond said that there "is no reason why it may not soon occur to a recording company that it can reduce the entire catalog of its offerings to a magnetic disk storage device. It could then offer its subscribers the capacity via their phone company line to dial (or key in) the catalog number of a desired composition— or a track within a complete work." Then, a computer could search the magnetic disk at random access speed to locate the work, read it out on a teleprocessing unit, and transmit it directly to a receiving instrument. Importantly, the computer would also debit one's account at the bank and credit it to the rightsholder "for the pleasure (and profit) the music provided."[38] In 1981 the *Los Angeles Times* profiled Digital Music Co., a new company started by telecommunications entrepreneur William von Meister. This venture aimed to send music digitally from an LA-based studio via satellite to cable TV subscribers, although in its initial stages it required music industry acceptance. The idea was that subscribers could make their own recordings from "static-free transmissions," bypassing the record store.[39] The goal of the service, called Home Music Store, was to allow a subscriber to use a decoder to select music from a catalog via the telephone, and then a credit card would be billed automatically. Many observers were suspicious of the idea, noting that there were problems relating to royalties.[40] A decade later, a satellite radio station called Rock Shop Radio launched in Europe, offering mail-order CDs, concert tickets, and electrical equipment at discounted prices. Listeners could order any CD played on air by phone or mail. Users with a satellite dish or cable access in the UK could pick up the service, with the mail-order component aimed at listeners in Eastern Europe, where CDs and devices like Walkmans were scarce.[41] Another example from the early 1990s that exhibited a partnership between digital radio and cable TV was one backed by Sony and Warner, called Music Choice Europe in Europe and Digital Cable Radio in America, which broadcast thirty-four channels, "as diverse as classical, soul, country, indie, and blues," directly to the home from a US satellite uplink.[42] Stingray Digital acquired Music Choice Europe in 2011. In these examples, attempts were made to automate the consumption process and circumvent the need to leave the home, visit a store, and purchase recorded music.

The notorious CD club, like those offered by Columbia House and BMG, was another service that employed a subscription-like membership model. These clubs were mail-order music services with a significant market presence that existed for decades before Sirius and XM. One difference between the clubs and most other music subscription services is that members owned and kept the records that were sent to them in the mail. But they still featured the ability to have music sent to the private home, and they tethered the consumer to the CD club.

Columbia House started in 1955 as the Columbia Record Club, a means for record label Columbia to sell music through the mail. Over the years, it added formats of the day, such as 8-tracks, cassettes, and CDs. The club hooked consumers by promising a number of albums for a penny: eight CDs or eleven or twelve cassettes, which increased to thirteen CDs and then went down to twelve. The catch was that one would then need to purchase more selections at the regular price to deliver on the membership agreement, the full terms of which few members read.[43] Rival BMG, of Bertelsmann Music Group, had typically advertised twelve CDs for the price of one.[44] While these clubs were in operation before the 1990s, BMG and Columbia House peaked in revenues during that decade. Columbia House hit $1.4 billion in revenue in 1996 before waning in relevance as the decade passed.[45]

Consumers who joined these clubs were sent a catalog every month, sometimes more frequently, and for each catalog one would need to send back a postcard within ten days to say that they did not want the selection of the month. If this was not done in time, the record of the month would be shipped, and the member would have to pay for it.[46] With membership, then, monthly shipments would be automatically sent unless one intervened. This is called negative option billing. Although the record clubs functioned as a membership, and were branded as such, the serial commitment and payment scheme, as well as the work it would take to remove oneself from the membership, mirrors the subscription models that are common today. If you've ever canceled a subscription service, or tried to, this likely sounds familiar. One fitting example is Amazon referring to its Prime cancellation process as "the Iliad," given the long and complex journey one faces when trying to cancel their subscription.[47] Youth, eager to build a CD collection, often came up against their parents, who were more wary of the membership details and the likelihood of having a difficult time canceling the subscription. In 1994, 15 percent of all CDs sold in the US came through these record clubs; three million of the thirteen million copies sold of Hootie & the Blowfish's *Cracked Rear View* were sold through the clubs.[48] As of 2000, the two companies were grossing $1.5 billion a year.[49] What also enabled the clubs to be profitable throughout the CD era was that they licensed master tapes and production files from the major music companies and pressed their own at a cost of about

$1.50 each. The clubs could then make about $5 on each album sold.[50] Further, the clubs, without written licenses to distribute the records, paid most publishers 75 percent of the standard royalties set by copyright law. Because the publishers were accepting these payments, the clubs argued they were submitting to "implied" licenses.[51] If a publisher or label complained, the clubs would remove their records from the catalogs.

There were notable responses to the exploitative practices employed by these record clubs. Over the course of their existence, these companies generated countless complaints from consumers due to questionable business practices. These included changing fine-print rules without informing subscribers.[52] In Canada, major retail chains protested the cosponsorship of the 25th Juno Awards by Columbia House, due to the effects of the clubs on retail sales. In 1996 it was estimated that the clubs accounted for 30 percent of all record sales in Canada, an estimated $200 million CAD. In 1995 MCA Music Entertainment Canada, and in America, Virgin Records, MCA Records, and Geffen Records, all pulled out of the clubs. Some industry observers went so far as to say that record clubs diminished the value of records and "cheapened" the "art form." One asked, "How can an artist be worth a $25–$30 [concert] ticket if 11 CDs are worth a penny?"[53] Class-action lawsuits were filed against the record clubs in 2003, estimating that the companies behind the clubs could be taking upward of $100 million a year in royalty payments owed to composers because the clubs were only paying 75 percent of the statutory rate.[54]

In 1999 the record clubs, perhaps in anticipation of increasing internet use, music piracy, and digital downloading, attempted to revise their strategies. One such strategy was Play, a club created by Columbia House that offered a "hassle-free" membership, as opposed to the negative option tactic. Play members had two years to choose at least six titles for purchase without any album shipping automatically.[55] Columbia House made efforts to embrace online retail, and in early 2000 it reorganized into three divisions, one being an online version of the music and video club. It hoped to reduce costs, particularly direct marketing through the mail, which relied on expensive printed materials. This allowed the company to list its sixteen thousand music titles online, a feat that was impossible in print. Columbia House, a 50–50 joint venture between Sony Music Entertainment and Warner Music Group as of 1991, had planned to merge with online retailer CDnow.[56] The planned merger was not pursued, with *Billboard* reporting that it was likely due to declining interest from Columbia House co-owners Sony and Time Warner as well as a weaker-than-expected financial situation.[57] Instead, Sony and Warner invested $51 million in CDnow. In 2002 the private investment firm Blackstone Group was negotiating to buy a majority stake in Columbia House;[58] it did so, receiving an 85 percent stake for $410 million.[59]

Columbia House and BMG merged in 2005, with BMG acquiring its competitor from the Blackstone Group, Sony, and Time Warner at a price of $400 million.[60] The two had been in negotiations in 2001 about a possible merger, but at that time the companies downsized and restructured instead. Columbia House closed two warehouses in Bloomington, Indiana, and Colorado City, Colorado, resulting in the layoffs of four hundred warehouse workers as well as one hundred at the company's headquarters. BMG laid off sixty-five people.[61] In 2010 Columbia House moved to a DVD-only model but struggled to compete in a media landscape increasingly dominated by streaming services launched by Big Tech companies that were better positioned to meet the demands of consumers. The mail-order CD club was declared dead by *Slate* in summer 2015 after the owners of Columbia House filed for bankruptcy.[62] At this time, the club still had 110,000 members paying for mail-order DVDs, which Columbia House obtained through licensing agreements with film studios. Many of these studios had become creditors, with Universal having been owed more than $1 million and Warner Bros. nearly $300,000.[63] But even in 2008, and likely before, the writing was on the wall for the record clubs. In July of that year, private equity firm Najafi acquired Direct Group North America, which included the Columbia House assets and the BMG clubs. Even before the sale of Direct Group, Bertelsmann, which owned BMG, said it would discontinue CD clubs in 2010. Interestingly, Ed Christman of *Billboard* pondered, "Who can say that they won't be back as a digital subscription service?"[64]

With Columbia House's subscription-like elements, business strategies that skirted royalty payments, and claims of record clubs devaluing music, it's fitting that *The Verge* called it "the Spotify of the '80s."[65] Spotify has faced critiques of its low payout to artists and tendency to privilege superstars in its payout systems, highlighting the fact that subscription models have for the most part failed to adequately compensate artists. What is also intriguing about the story of the CD clubs in the 1990s is that Sirius and XM launched in the years immediately following the height of the record clubs' success. The satellite radio services not only had to convince listeners who might be skeptical of long-term commitments in the form of subscriptions for music at a time when ownership was still largely dominant in terms of how people collected and listened to music, whether CDs or MP3s, but they also had to convince listeners that radio, a medium that is often assumed to be free, was worth paying for.

TV ON THE RADIO

Before satellite radio, the television industry had established the logic of subscription access for exclusive programming. In the wake of the Sirius manifesto ad campaign mentioned earlier, Sirius VP of programming and market development Larry Rebich explained, "We're not really radio, in the sense that HBO or Showtime are not TV."[66] The 1975 satellite debut of HBO was dubbed a "revolution in cable programming" because it was the first time a non-broadcast-based cable network was available across the US.[67] HBO and other users of satellite and cable were styled in revolutionary and utopian rhetoric, as services that could remedy the "perceived ills of broadcast television, including lowest-common denominator programming." The period between the late 1960s and early 1970s is called cable's "Blue Sky" period for this reason.[68] Cable TV offered programming beyond "sanitized network content" in a similar way that satellite radio was to be the home for listeners "who enjoy broader options."[69] Just as the satellite positioned Sirius and XM against commercial broadcasting, it had accomplished something similar for the medium of television.[70]

Sirius and XM sought advice from cable and satellite television companies in launching and refining aspects of their services, particularly with regard to how to manage relations with subscribers. The satellite and cable television industry was a fitting place to find insight into developing a service with satellite infrastructure and national reach. In 1999 XM established a consulting services agreement with DirecTV to provide professionals to assist XM in establishing a customer care center and billing operations. This partnership ended in January 2004.[71] Statistics pertaining to cable and satellite television also provided the radio companies with some level of confidence in whether or not listeners would pay for premium programming. XM explained that data relating to cable, satellite TV, and premium movie channels confirm that consumers will pay for services that expand choice or enhance quality. As of 1999, 67 percent of American households subscribed to basic cable television and 11 percent to satellite television, at an average monthly cost of $51. In the same year, more than seventy-five million subscribers to cable and satellite services made the additional purchase of a premium channel unit, like HBO.[72]

One other intriguing parallel between radio and screen media is the way that ideas of scale and expansion were applied to generate interest in viewers and listeners. As Charles R. Acland explains with respect to cinematic technologies of the 1950s, the various widescreen and large-gauge formats served a purpose of showcasing distinctive technological features of the films that employed them.[73] An example of this that connects to the satellite radio universe is Paul McCartney's *Oobu Joobu* radio show, described by McCartney as "widescreen radio." The show aired in 1995 on American network Westwood One in the years leading up to the launch of satellite radio. As such, it's an interesting proto example of a sort of grandiose radio, in which the technology of visual media, in this case the widescreen, comes to indicate a type of radio *beyond* radio. Fifteen radio shows included clips of demos and unreleased material by McCartney and The Beatles, "from Paul's own private collection"; never-before-heard recordings; clips and songs by other musicians; and chats with guest artists like Stevie Wonder.[74] A number of *Oobu Joobu* clips are available on YouTube; one simply titled "Oobu Joobu Pt.3" provides a sense of the "widescreen" aspect ratio of the program. It begins with McCartney introducing himself and saying, "You're listening to *Oobu Joobu* and it's wide screen radio" (with emphatic pauses between "wide," "screen," and "radio"). He then repeats, "It's wide ... screen ... radio," and with each syllable the sound transfers between the left and right channels to present a sense of expansive space.[75] There are also moments when McCartney's voice cuts in with an echoing reverb effect. The Beatles Channel on SiriusXM regularly airs clips from *Oobu Joobu*, and the format, rare and exclusive access to a major celebrity's personal music collection, influences a number of

Sirius and XM channels and programs, such as Bob Dylan's *Theme Time Radio Hour*.[76]

Satellites also generated ideas about expanded programming and audiences. With their enhanced reach and ability to transmit multiple channels or programs, satellites could court those who had not been represented or reflected in network broadcast programming. As one example, satellites were said to be having an effect on Black contemporary radio, with the 1983 airing of the Sheridan Broadcasting Network. All of the network's programming was Black-oriented and offered as dayparts so affiliates could carry services as they like. Programming included news and information, as well as a night service that played music from contemporary jazz and rhythm and blues.[77] In XM's own descriptions of its programming, a similar emphasis on underserved listeners was apparent, something the company felt spoke well to the nature of its ability to advertise nationally to niche but geographically disparate groups. It cited the fact that it had "channels devoted to urban formats, the nation's first African American talk channel, five channels devoted to Latin music, as well as CNN *en Español*, three comedy channels and two channels devoted to kids and parents."[78] Thus, one of the perceived advantages of cable and satellite television is the technical capability to deliver more channels, which should, ideally, mean that a much greater variety of program options is available to a wider variety of people. *Narrowcasting*, the targeting of niche audiences, is one of the major changes to television that began in the 1990s. It followed significant growth in cable systems, from around 2,490 in 1970 to more than 10,000 in 1999, but this growth did not make cable access uniformly available. In the early 2000s, cable had reached nearly 70 percent of US homes, and the greater percentage of those without cable were concentrated in African American neighborhoods.[79] A similar disparity existed in internet access of that era (and still does). This disparity between those with access, and those without, must be kept in mind when considering how narrowcasting targets specific viewers or listeners. Television scholar Beretta E. Smith-Shomade argues that narrowcasting "encourages a center—a space where the really important demographics reside."[80] Further, as communication scholar Megan Mullen outlines, most early satellite networks relied on familiar program genres, those already proven successful on broadcast television.

Due to the significant expense of uplinking to satellites, "the additional cost of instituting major new programming infrastructures would have put most of them out of business."[81] Even more recently with streaming media, these revolutionary claims are mapped onto ideas about how the technology unquestionably improves diversity and representation. Media scholar Kristen Warner challenges this ongoing technological determinism to encourage us to "think harder about the PR and brand-friendly notion that streaming networks provide actual, meaningful democratization of content."[82] In the multiple channels of cable or satellite TV and radio, it's the affluent consumer who is the primary target.

Echoing these points made by Smith-Shomade, Mullen, and Warner, Alexander Russo and Bill Kirkpatrick argue that although satellite radio has created more channels, "it did nothing to alter [the] industrial conception of audiences as taste communities at whom radio pushed content."[83] The audiences that gravitated to Sirius and XM were the same already sought-after audiences of broadcast media, but with satellite radio, the more wealthy and mobile (or *auto*mobile) listeners were targeted. That said, the expansion of radio formats with smaller niche music channels on satellite radio has effectively reached subscribers by affirming their tastes and offering a home for music not regularly heard on commercial radio. Arguably, satellite radio aims to strike a balance between the familiar and the surprising in order to grow its subscriber base.

THE PURSUIT OF SATELLITE RADIO SUBSCRIBERS

One of the most central questions discussed about Sirius and XM throughout the 2000s was whether or not the companies were successful in attracting subscribers.[84] Pursuing subscriber-listeners was integral to both companies' business models, as this was where the vast majority of revenues were realized. In 2007 XM indicated that the company counted radios individually as subscribers.[85] Analysts had predicted that a critical mass for the "public acceptance of the technology" would occur when subscriptions to both XM and Sirius totaled a million. Just weeks before its two-year anniversary on November 12, 2003, XM alone hit the million subscriber mark.[86] Radios that were installed in vehicles and those for sale as portable devices in electronics stores generated the figures contributing to subscriber numbers, although the tastes, personalities, and demographic profiles of the people purchasing the devices and choosing to subscribe would be what influenced marketing campaigns and programming initiatives.

Both XM and Sirius relied heavily on the space of the car as a key battleground for establishing their services. In 2000 Sirius CEO David Margolese said that Sirius was "the pipe into the car" and that despite the satellite service planning to be available on portable radios, "the killer application remain[ed] the car."[87] Margolese had also said that the "key value driver" for the business would be the ubiquitous adoption of satellite radio by automakers, and he wanted its radio receivers to be "as common as airbags in cars."[88] The ability to have satellite radios be factory installed in vehicles played a major role in showcasing the service and in enabling a seamless transition from a pre-installed free trial to regular monthly payments. Having an available trial for listeners is one way to market intangible music, as it demonstrates the service's attributes and ideally leads the listener to come to rely on the service and then justify a subscription.[89] In October 2003 it was reported that in-car listening, as a percentage of total radio listening, had increased to 34 percent from 30 percent over the previous five years. Listeners surveyed said that interest in satellite radio grew with the option of having pre-installed radio units in their cars.[90] Both companies were active in securing deals with major automobile manufacturers as they competed for subscribers. In 2005 Ford announced that it would offer Sirius as a factory-installed option in more than twenty-one models and expressed the goal of adding one million new subscribers to Sirius in a period of two months.[91] In December 2004 General Motors signed its one-millionth XM subscriber, and one in three GM vehicles was equipped with a factory-installed XM radio.[92] To encourage the broad installation of radios in vehicles, the companies subsidized their installation. XM subsidized a portion of the cost and made incentive payments to GM when car owners became subscribers. XM shared a percentage of subscription revenue attributable to GM vehicles.[93] XM had also secured Toyota as a sole supplier of factory-installed radios as well as Hyundai, and its radios were installed in Honda and Acura vehicles.[94] XM saw original equipment manufacturer (OEM) subscribers increase

73 percent in 2005 compared to 2004, with XM becoming available in more than 130 vehicle models. Approximately 120 models offered factory-installed options. Factory-installed options were added to more than 40 vehicle models in 2005.[95] XM noted in 2005 that the company was retaining approximately 60 percent of customers who received a promotion subscription when purchasing or leasing a new vehicle.[96] The companies also monitored subscriber turnover, or churn. In 2005 the average monthly churn (the average of the number of deactivations divided by average quarterly subscribers) was 1.5 percent.[97] The aim was to convert free trials into subscriptions and to do so through retention and win-back programs and customer service strategies.[98] As of December 2006, Sirius radios were available as a factory-installed option in 132 models.[99] Both companies were intensely focused on in-vehicle expansion and becoming available across a range of vehicle models.

One major development in Sirius's early years was its monumental deal with "shock jock" Howard Stern. The deal was announced in 2004, but the show only went live in 2006. Acquiring Stern established a higher level of exclusivity for Sirius, with a more uncensored and uninhibited listening experience becoming available to subscribers. Stern said that he was "creatively shackled by the FCC," and he had been threatening to move to satellite radio for most of 2004.[100] In that year, he signed a five-year, $500 million deal with Sirius to start on January 1, 2006. The Stern deal was distinct from the statutory licenses that set royalty rates for music.[101] It was a marketplace rate, meaning that it could be above and beyond the payouts that music rights holders receive for airplay on satellite radio.[102] Because of the cost of Stern's show, including salaries and construction for a studio, a cost estimated at $100 million per year, he would need to generate about 1 million subscribers paying the monthly fee of $12.95 to cover the costs of the deal. Sirius, as of October 2004, had around 600,000 subscribers, much less than XM's 2.1 million.[103] By early 2005 Stern's influence was apparent, with Sirius's stock price more than doubling in 2004, a gain of 141 percent.[104] By April 2006, Sirius outpaced XM in subscriber growth for the first time.[105] A poll conducted a few months later reported that 32 percent of those surveyed had listed Stern as the key factor in their decision to subscribe.[106] Stern was paid a bonus of nearly $83 million in early 2007 for surpassing the subscriber goals set in 2004, helping to take

Figure 10. XM's partners. *Source:* XM Satellite Radio, *2006 XM Annual Report*, 2007.

Sirius from 600,000 subscribers when it signed Stern to over 6 million at the end of 2006.[107]

Between 2004 and 2007, both XM and Sirius, under CEOs Hugh Panero and Mel Karmazin, respectively, were busy promising new program offerings that the companies hoped would retain current subscribers and entice new ones.[108] In Karmazin's letter to stockholders in 2005 he said, "Content is 'king' in the entertainment industry, and SIRIUS will continue to rule this space in radio."[109] Musicians such as Snoop Dogg, Eminem, "Little Steven" Van Zandt, and Tom Petty were among the first to host or produce shows on satellite radio. Like the deal with Stern, the Eminem deal would be profiled as one well-suited to a "safe haven for controversial artists." As Eminem said, "No middleman, no playlists, no bullshit. And most of all, no censorship."[110] Sirius paid $220 million for five years of the NFL and $107.5 million for five years of NASCAR, starting in 2007. When Sirius announced the NASCAR partnership, it was said that NASCAR had seventy-five million fans, who purchased over $2 billion in licensed products annually, and that "SIRIUS [was] going to tap into this market."[111] XM had its exclusive sports coverage as well, including NASCAR before Sirius, but also paid $650 million for eleven years of MLB baseball.[112] These acquisitions helped secure subscribers but were also massive financial investments, on top of music royalty payments, which would increase with subscriber growth. Making reference to Sirius's expensive deal with the NFL, XM's 2005 annual report explained that the company might not be able to obtain similar third-party programming if these massive deals made it more costly to obtain.[113] One analyst reframed the question "Can you make subscribers out of radio listeners?" to now ask whether subscriber growth could make up for content costs.[114] Sirius saw its programming and content costs more than double in 2004, from $30,214,000 to $63,949,000,[115] while subscriber revenue grew from $12,615,000 to $62,881,000 over the same period.[116] Other methods of driving subscriber growth were also costly. Subscriber acquisition costs included subsidies paid to radio manufacturers, automakers, and chip set manufacturers, as well as commissions paid to retailers and automakers. The majority of these costs were incurred in advance of acquiring a subscriber.[117] In May 2006 it was reported that Sirius's first-quarter financials showed growing losses due to spending for the purpose of growing its subscriber base. Revenue had jumped to $126.7 million from $43.2 million a year earlier, with a net loss widening to $458.5 million from $193.6 million the year before.[118] Throughout 2005 both companies noted quarterly losses. By early 2007 reports emerged indicating that Sirius wanted to merge with XM, and a merger deal was struck on February 19, 2007.[119] This was in spite of Sirius having set an annual record for new subscribers in 2006, 2.7 million, which gave it a total of over 6 million.[120]

A merger would enable the companies to share satellites and programming, among other things. They also would not have to compete against one another in a music market that was full of new services and devices that made use of digital technology and the internet. Karmazin also faced the pressing question of what it would take to keep Howard Stern with Sirius. In 2008 it was reported that both XM and Sirius had "burned through billions in the past decade."[121] The merger was delayed as regulators considered whether the merged company would constitute a monopoly or one company in competition with internet radio, traditional radio, iPods, and so forth. Ultimately, the merger received FCC approval July 25, 2008.[122]

AUTOMOBILITY AND MOBILE PRIVATIZATION

In the early years of Sirius and XM, a particular type of listener characterized the ideal subscriber: one who was intrigued by new technologies and had the purchasing power to cover a subscription, as well as an automobile equipped with satellite radio. Thus, a major implication of the shift from *listener* to *subscriber* is the hierarchy it exemplifies. A subscriber typically imagines their taste in music as not being served by regular commercial radio fare. As XM proudly boasted in 2002, the company "creates fans—not just listeners. Our world-class programming team and programming partners are indeed all about treating music listeners like music lovers . . . because loving music is what satellite radio is all about."[123] There are both class and cultural hierarchies implicated in the subscriber-listener. Sirius and XM imagined a characterization of the ideal subscriber, one that shifted and changed as the companies sought out more subscribers. XM explained in 2000 that its audience was younger and more educated than network radio listeners.[124] In 2002,

40 percent of subscribers were age thirty-four or younger.[125] In 2003 executive VP of marketing for Sirius Mary Pat Ryan said that Sirius's audience was "people who buy 20 or more CDs a year, go to concerts and subscribe to music magazines."[126] At this time, XM also pointed to the over-thirty music listener as ideal because "there are more passionate music fans over the age of 30 than ever before," with chief programming officer Lee Abrams citing the fact that they grew up in the "musically rich climate of the '60s and '70s."[127] At the end of the decade, following the merger of Sirius and XM, an Arbitron study highlighted some key aspects of SiriusXM's thirty-five million total adult listeners. On a typical day in 2009, SiriusXM listeners spent two hours and forty-five minutes in their vehicles, and 71 percent of their in-vehicle listening was to SiriusXM. The study revealed that SiriusXM listeners indexed high on education, income, and receptiveness. Of all listeners, 56 percent had graduated from college or had advanced degrees, compared to 24 percent of AM/FM listeners, and 24 percent had household incomes of greater than $150,000, compared to 9 percent of AM/FM listeners.[128]

When thinking about spaces of listening, the notion of class and cultural hierarchies becomes even sharper. The automobile was an essential component of who subscribers were and still are. Not only is automobile ownership central to a satellite radio subscription, so too is the time spent in a vehicle. Ideal subscribers spend time commuting between the home and the office or the home and a cabin, cottage, or summer home. Radios have been a central part of the vehicle since the medium's golden age, but particularly throughout and after the 1950s, when the television became the dominant entertainment medium in the home. The radio became associated with youth culture and mobility, and even rebellion.[129] Driving can evoke freedom, and music becomes a soundtrack for the open road. This philosophy of freedom on the open road was visually apparent in Sirius's early annual reports. Its 2002 report, for example, includes a first-page image of a car driving with an open road ahead. On the horizon is a lens flare, suggesting that the destination is not what is important but rather the time spent in the vehicle. On the dashboard player we can see that REM's "Drive" is being played. Vehicles have been described as a new living room for mobile families with children, where time spent transporting kids from school to home and to extracurricular activities

Figure 11. Delphi XM SKYFi audio system. *Source:* XM Satellite Radio, *2002 Annual Report*, 2003.

becomes a significant component of one's everyday life, particularly in a context of suburban living. Beginning in the 1920s, marketing strategies for the automobile privileged interior comfort for middle-class families,[130] and throughout the 1930s the car radio was connected to domesticity.[131] In the late 1940s Ford was advertising certain models as the "living room on wheels."[132] The car, according to Jean Baudrillard, "makes it possible to be simultaneously at home and further and further away from home."[133] One Sirius product that was temporarily available echoes this sentiment. The

Figure 12. "XM is driving ahead" ad. *Source:* XM Satellite Radio, *1999 Annual Report*, 2000.

Backseat TV video service was announced in March 2007 and included streaming video from family TV channels like Disney.

The vehicle as a mobile, private space for listening to radio is a notion reflected in Williams's explanation of mobile privatization. Williams has said that the radio industry was a major sector of industrial production by the end of the 1920s, part of a "rapid general expansion of the new kinds of machines which were eventually to be called 'consumer durables.'" This complex, he added, was characterized by two paradoxical but connected tendencies of modern urban industrial living: mobility and the self-sufficient family home.[134] This "at once mobile and home-centered way of living" was a form of mobile privatization, whereby the distances between home and places of work, and other relevance, increased. Automobiles have increasingly become "sophisticated mobile sound machines," but as early as the 1930s, American auto manufacturers recognized the desire to have radios in the car.[135] Between 1936 and 1941 over 30 percent of cars were equipped with radios.[136] The automobile is also a much more controlled space than other places where listening takes place, such as the living room, and it is possible for acoustic designers to "create a uniformly pleasant listening environment."[137] In one's own vehicle, they can be alone and customize space while traversing public places. In *Sound and Safe*, Karin Bijsterveld et al. explain that this "sonic capsule enables the phenomenon of acoustic cocooning," with a driver feeling relaxed by controlling the interior acoustic environment of the car.[138] Within this context, the emphasis on the vehicle for the acquisition of subscribers is an apt point for thinking about the cultural relevance of satellite radio in the 2000s and beyond. Following a stage of mass production and consumption of the automobile from 1925 to 1960 is a stage of diverse products and branding and lifestyle choices for car consumers, in which the added feature of a satellite radio receiver has become an appealing option for customizing the private, mobile, and sonic space of the vehicle. It enables the vehicle to be an object to express or complement a notion of individuality.[139] A radio service free from government regulation and that does not cut out as we drive away from a transmitter effectively aligns with this trajectory.

Car ownership and the experience of freedom through mobility points to hierarchies of class, especially when considering the demographics tied to car ownership, travel, and income for subscriptions. Connected to class hierarchies are questions of mobility and who is free to move at will and with the comfort of a vehicle. A "new mobilities paradigm" aims to make sense of a world increasingly on the move; the power and practices of mobility in creating movement and stasis; and the fluidity and movement that correspond to "extensive systems of immobility," like communications infrastructure such as satellites.[140] Sociologists Mimi Sheller and John Urry argue that time spent being on the move is not dead time.[141] Travel time for many subscribers of satellite radio, and radio more broadly, is crucial listening time.

In a critique of mobility as a new universalizing condition, one with an abject relationship to class and power structures, sociologist Beverley Skeggs expresses concern with the decline of class cultures in studies of middle-class individualization and individual identities, theories, and claims, which, Skeggs argues, institutionalize and reproduce class hierarchies.[142] But class and individuality are linked in the figure of the subscriber-listener, with class and structures producing the means by which a middle-class individuality can write about itself as becoming a universalizing concept (one of Skeggs's concerns). The middle-class driver with an ideology of individuality was an ideal subscriber-listener for Sirius and XM in their early pursuit of, and competition for, subscribers. Skeggs also points to Pierre Bourdieu's metaphors of capital as appropriate for understanding "who *can* move and who *cannot*, and what the mobile/fixed bodies require as resources to gain access to different spaces."[143] Bourdieu's sociological concepts that make sense of cultural hierarchies determined by social origin and ways of being are helpful in characterizing the subscriber-listener. While disposable income determines one's ability to subscribe, not only to one service but to many, one's cultural capital and even habitus are the next layer for thinking through why a music lover would subscribe to satellite radio. Sound studies scholar Jonathan Sterne also advocates for a consideration of Bourdieu's concepts in studies of technology to help overcome unhelpful binaries like technology/society. Technologies are socially shaped and socially located, Sterne says, along with meanings, functions, and use. Bourdieu's concept of habitus, "the way a person walks, talks, types, plays a musical instrument, drives, her aesthetic preferences, perceived health needs, etc.," is tied to the social use and transformations of technologies.[144] The reasons why some might be drawn to subscription radio call to attention the concept of habitus, along with forms of capital. For some, the regular use of a car is the status quo, as is the practice of moving between spaces of the home, work, leisure activities, vacations, and other forms of travel. Their lifestyle may also be shaped by the customization of consumer and communications technologies, such as the option to equip the car with satellite radio. A sense of exceptional or elevated taste is also apparent, particularly in the way the satellite services were marketed to listeners through exclusivity and niche styles of music. To subscribe and have access to a Deep Tracks channel is to exhibit taste that is distinct from the mass offerings of ad-supported commercial radio and to reflect or reinforce one's cultural capital.

The sort of cultural capital on display in the satellite radio universe, however, is not shaped by clear cultural binaries of high and low culture. For example, it's unlikely that a subscriber is only after one channel, such as the Met Opera channel, although its availability may be a major reason that one subscribes. More likely, a subscriber is drawn to the wide range of selections that can be explored and experienced with some fluidity. This idea reflects the cultural context of the new millennium. In the 1990s the high/low culture binary was complicated by a turn to consider taste as shaped by an ability to be well versed across these hierarchies, a sort of cultural omnivorousness. This concept accounts for "the tastes and consumption practices of privileged individuals and groups at the end of the twentieth century," in which a fluidity across genres and levels of cultural value comes to shape one's expertise across popular culture, such as the ability to simultaneously delineate the merits of both Carly Rae Jepsen and Chopin.[145] To return to the question "What's in your orbit?," the wide range of cultural touchstones one is familiar with and the more random or spontaneous these tastes seem exhibit a sort of cultural capital or expertise that pairs well in an era of noise and information overload. Who, then, is able to navigate this cultural landscape with authority, and to what services do they subscribe?

Beyond the vehicle, satellite radio also became available in other technologies of movement, like airplanes. AirTran Airways was the first to offer satellite radio as in-flight entertainment, beginning with three planes and expanding to twenty in February 2005. Marking this new in-flight service were twenty Boeing 717s that featured a custom design with the XM logo and Elton John, who was an XM spokesperson at that time.[146] Sirius radio receivers were also made available as a standard feature in all CD head unit–equipped boats built by Genmar Holdings Inc., the world's largest recreational boat builder.[147]

Among XM's initial subscribers, around 80 percent were located in major urban areas.[148] Terrestrial repeaters receive the satellite signal and amplify it for retransmission at a higher signal strength to overcome signal obstruction, an obvious issue in areas with tall buildings and bridges. As one journalist described the network in 2000, it is a "56-city system of earth-mounted repeaters," "a series of norad-like megadishes, visible only by helicopter" or from a higher building.[149] XM had a contract to purchase 1,550 repeaters, with a goal of installing as many as 1,700 in seventy markets, with major cities potentially requiring over 100 repeaters.[150] Terrestrial broadcasters were concerned with the repeater networks and the potential issue of satellite services competing locally in programming or interfering with radio signals. The NAB called on the FCC to take action against the satellite companies, calling the repeater networks "a crutch for a technology that is not up to the task of providing seamless, mobile coverage," worried that the repeaters were designed to "blanket cities rather than fill in gaps." XM, for instance, had proposed a total of sixty-six high-powered repeaters for the city of Boston.[151] Despite the NAB's frustration, the repeaters continued to buttress satellite signals.

Within the space of the city, a different sort of mobility is present, beyond that of the space of the car. In cities are people on the go, in and out of office buildings, apartments, coffee shops, and hotels, and in transit via major transportation hubs. The city brings density, and with it, more potential subscribers. As urban studies scholar Stephen Graham writes about the vertical makeup of the city, the height of towers is associated with wealth and value. Towers are "stage sets for media stunts designed to lubricate the worlds of tourism or hyperconsumption."[152] Like a skyscraper standing out from the rest of the city, the subscriber, in the satellite's view, stands out from a mass of listeners.

Throughout the 2000s, any major city in North America was home to multiple Starbucks locations. In 2004 XM announced a multiyear strategic marketing partnership with Starbucks.[153] This partnership included the creation of the Starbucks Hear Music channel, which featured music from the coffee company's Hear Music label. More than four thousand American Starbucks locations played the channel, exposing it to countless potential subscribers who were also Starbucks customers. Artists who contributed music to the XM Hear Music Series included Jason Mraz, who had three non-album songs on a compilation album released in stores, as well as Tracy Chapman and Jewel.[154] Starbucks expanded the Hear Music concept into the Hear Music media bar, an in-store kiosk that allowed customers to buy customizable CDs on demand. There were also Hear Music Coffeehouses, which were coffee shops stocked with fifteen thousand CDs in addition to digital music offerings. In October 2004 a total of forty-five stores in Austin and Seattle were expected to be equipped with kiosks. Customers could use tablet PCs to browse and purchase seven songs for $8.99, choosing from 150,000 tracks that came from content deals with major and independent labels.[155] By 2007 synergies were in full effect as to promote artists receiving frequent airplay on the Starbucks channel and whose work was for sale in the coffee shops themselves. The band Low Stars, said to be Starbucks's first previously unreleased act, had an album out on the Hear Music CD Series. Starbucks had hoped to act as a label in developing the band as well as a retailer, by stocking an estimated 60,000 to 100,000 CDs in its stores. Promotions for the album included the CD being featured at the checkout counter, in-store airplay, and programming on the XM channel.[156] These promotional partnerships allowed XM to use the widely familiar Starbucks brand to qualify the satellite service as something familiar and well-suited to the burgeoning coffee shop culture of the decade. However, a reliance on the CD as a key object in the promotional strategy signaled the limited term of the initiative. In December 2007 XM signed a termination and release agreement with Starbucks, issuing the coffee company nearly two million shares of its Class A common stock.[157] Today, SiriusXM's own Coffee Shop Radio occupies one of the prime top 10 channels on the satellite dial, indicating the continued targeting of coffee shop musical ambience in the channel lineup.

In addition to the coffee shop, the satellite companies also targeted luxury hotels as places where potential subscribers might be found, pointing, again, to the sort of lifestyle of the ideal subscriber these companies had in mind: one who travels; can afford higher end hotels; and might find it satisfying to have access to the same radio service in the car, the airplane, and the hotel. In 2005 W Hotels in New York and Los Angeles were outfitted with Sirius radios as an initial test to see if a more substantial rollout would be justifiable.[158] The W, as a hotel that caters to younger customers, and the setting of the major American cities of New York and Los Angeles, provides further insight into the type of subscriber Sirius was targeting at this time. In the same year, XM initiated a multiyear deal with Hyatt hotels to install tabletop radios in more than fifty thousand hotel rooms in the US.[159]

These brand partnerships also reflect efforts to seek out ideal subscribers in a range of targeted places and spaces. But the listening practices of people are regularly transformed by innovation and change.[160] This means that these key places and spaces also change, especially as new technologies determine where and how we listen. According to an Arbitron study, thirty million people listened to internet radio each week in January 2004 in the US.[161] The increasing desire to take music with us shaped how satellite radio developed. In 2005 XM spent considerable time in its annual report detailing the Apple iPod as a significant competitor to its services, whereas earlier reports had tended to focus more on the availability of free radio programming from commercial broadcasters. In 2004 Apple sold over 4.4 million iPods, a device that was also compatible with certain car stereos and home speakers.[162] Sales of online music in both the US and Europe also increased tenfold between 2003 and 2004.[163] In Sirius's annual reports as of 2005 the ability of subscribers to access internet streams of music channels and selected non-music channels was regularly emphasized, and the company's online features would expand in future years.

In 2005 internet service provider AOL teamed up with XM to launch an online radio service that provided 70 XM channels and AOL's 130 streamed channels to AOL subscribers for free. XM would include Radio AOL, AOL Music Sessions, and AOL Music Live on the satellite service. CEO Hugh Panero reinforced XM's key strategy at this time, explaining that the company's "philosophy has always been to work with strong partners at every level: retail, automotive, and now online to build awareness and subscriptions for XM."[164] XM had also turned to record industry foe Napster to create a digital music store and subscription service. Napster, having been forced to rethink its role in the wake of having its peer-to-peer

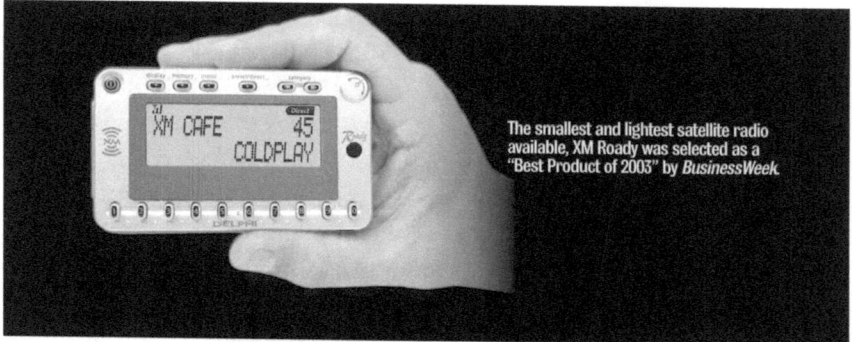

Figure 13. XM Roady ad. *Source:* XM Satellite Radio, *2003 Annual Report*, 2004.

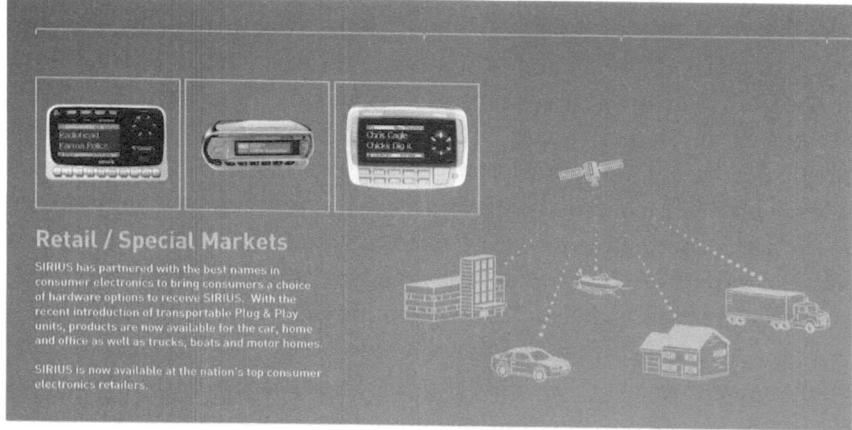

Figure 14. Sirius's plug & play units. *Source:* Sirius Satellite Radio, *2002 Annual Report*, 2003.

file MP3 exchange system shut down in 2001, found a willing partner in XM as the company looked to find ways to use the internet to its advantage. The store and service, dubbed XM + Napster, used a new line of MP3 players, such as the Pioneer Inno, to access XM programming and store up to fifty hours of XM feeds; when connected to a computer, a user could purchase or save tracks from the Napster to Go subscription service.[165] The Roady XT, introduced in August 2005, included a built-in FM transmitter that could make any FM radio an XM radio by wirelessly transmitting to any one of its one hundred FM radio frequencies. It came with a customizable sports and stock ticker and a TuneSelect feature that would alert the listener when a favorite song or artist was being played on any XM channel.[166] Sirius radios had a similar function, called SIRIUS-Seek, and many of those radios allowed listeners to pause, rewind, and fast forward.

Sirius also established partnerships to help break into the wireless and internet radio environment. Wireless operator Sprint debuted a radio service with Sirius called Sirius Music in 2005, which gave Sprint subscribers access to twenty commercial-free channels. The programming did not come through the satellite network but through the wireless cellular network. Sprint canceled the Sirius partnership in early 2007 and replaced it with ten genre-based channels of music from another provider.[167] The Stiletto 100 was Sirius's own portable radio receiver, which doubled as an MP3 player. An improvement on the early model, the S50, it could record and store songs from the radio stream.[168] The Stiletto 100 could store up to one hundred hours of broadcast music or ten hours of individual songs. Individual songs could be flagged, or "loved," to later be purchased from Yahoo. If no satellite signal was available, the radio would use a Wi-Fi connection to continue streaming from the internet. To no one's surprise, given the larger context of the record industry fighting against any device or service that would allow a user to store or save music without paying for a record or an individual song, the RIAA argued that this sort of "platform convergence," which had hard drives combining with broadcasts to create download services, would require content protection to prevent users from redistributing personal libraries without compensating rightsholders.[169]

These efforts to break into internet radio by Sirius and XM indicate that the companies fully realized the need to adapt and appease listeners who were increasingly online. Internet radio allowed for a more seamless transition between spaces of the home, the car, and elsewhere. Efforts to partner with existing companies in an increasingly uncertain media landscape were largely unfocused and unproductive, not to mention difficult to accomplish without clear support from record labels. Throughout the early years of Sirius and XM launching and operating their radio services, subscribers were integral to their growth and subsequent goals. The need to expand, to find new listeners in new places and spaces, would continue to shape business strategies post-merger. This reinforces Jody Berland's argument that the "cultural and geographic expansion of audiences is necessary for the economic expansion of the cultural industries."[170] As music media change from mass to niche, listener to subscriber, they become more personalized, focused, and intense, as well as less collective, random, and varied. After Sirius and XM merged, the efforts to develop usable and effective online aspects of the radio service intensified, aiming to weather new competitors in the radio and music industries, like Spotify, Apple Music, and the devices that would enable these services to enter the vehicle.

Satellite radio requires subscribers to operate. Its music programming helps turn listeners into subscribers, but programming and content costs are significant. To entice subscribers, narratives and discourse in the trade press, annual reports, and advertisements used the novelty of satellite radio to promote exceptionality with respect to what satellite radio could offer listeners, particularly against commercial broadcast radio. But as media scholar Thomas Streeter has argued with respect to cable and satellite television, cable "has not revolutionized the basic corporate structure of television. It has been integrated within it."[171] As with cable, the revolutionary aspect of satellite radio was an expansion of bandwidth for which to offer a greater number of channels. Satellite radio expands music programming in a way that establishes close connections between music and listeners, or rather subscribers, across a range of music styles, which remains part of its appeal. Programming is also a means by which Sirius and XM, and later SiriusXM, have developed and maintained musical elements that communicate perceived value to listeners. The next chapter explores this process in greater detail.

3 Perceived Value and Satellite Streams in Music Programming

Music as a *stream* is a common metaphor. Today, listeners regularly encounter recorded music via a streaming platform rather than physical media. Music is delivered in small amounts of data that, with a reliable internet connection, provide uninterrupted sound. Sound quality is of varying bitrate, or quality, depending on the service used or user preference and, potentially, how much one spends on a subscription.[1] And the metaphor has a longer history. Before the advent of streaming titans like Spotify and Apple Music, Sirius Satellite Radio used the word *stream* to refer to channels beamed down by satellite to one's radio receiver. As a foil to the notion of a station, something grounded in place, a stream is more ethereal and suggests fluidity and movement between outer space and Earth. On-demand streaming music is an idea that goes back over one hundred years. According to scholar and critic Eric Harvey, it can be traced back to the science fiction novel *Looking Backward* (1888), which featured a music room with twenty-four-hour music playlists for subscribers.[2] Jeremy Morris and Devon Powers investigate the ways that music streaming providers work to ensure a "valuable experience" for users in an environment of endless choice by highlighting how streaming serves as a metaphor for unlimited access to content and the flow of information. They explain that streaming has never held a single meaning or practice but "implies freedom and bounty as well as limits and constraints."[3] Hendrik Storstein Spilker and Terje Colbjørnsen argue that it's an evolving concept with many manifestations over the years.[4] The term's mobility, fluidity, and flow make it apt for describing the music channels offered by Sirius and XM, which move up and down the lineup and change over time.

In much of XM Radio's early promotional materials, the company said that it "beams 101 channels of music, news, sports and talk programming to customers coast-to-coast."[5] A stream is not unlike a beam. Both are fluid and constant and from the beyond. Both can potentially engulf a person. Sirius would abandon the use of *stream* as the years went by, adopting the more familiar term *channel*, but the notion of the vast array of program options streaming or beaming down from space tells us much about how value, or the worthwhileness of paying for subscription access to music radio, was crafted by the satellite radio companies. Music is an essential feature of Sirius's and XM's program offerings, and it is used to communicate and advertise the perceived value of the subscription services. Music's emotional and subjective qualities connect with listeners and their sense of self and identity. This chapter focuses on the ways that Sirius and XM have crafted music programming on satellite radio, showcasing how music serves to justify subscription access to radio. Satellite radio streams, or channels, have in some ways been modeled on broadcast radio formats, but the expansive channel lineup allows for the celebration of niche or neglected radio formats and for 24/7 artist-based channels. Perceived value is a concept that accounts for the intangible feeling of exclusiveness that subscribers may have, a feeling inspired by the wide variety of channels available to subscribers as well as the on-air celebrities and radio personalities who bring music to life.

Sirius defined its music streams as each being operated as its own individual radio station, with a distinct format and its own hosts. The company began with sixty music streams, which enabled it to "superserve" subscribers with a range of content that surpassed what AM and FM radio offered listeners. Sirius called its program managers "stream jockeys," and most came from the recording, broadcasting, and entertainment industries to manage the daily program for each music stream.[6] Staff members espoused their pleasure at their freedom from commercial

Figure 15. Sirius's 100 streams of satellite radio. *Source:* Sirius Satellite Radio, *2002 Annual Report*, 2003.

broadcast radio restrictions and limited or repetitive playlists. Streams were broken down into genre categories. For example, there were six dance streams (house music, nonstop club mix, mainstream dance, electronica, dance hits, disco) and a series of decades streams that fell under the umbrella of pop (1950s, 1960s, 1970s, and 1980s). Rock had the most streams, thirteen (including blues, indie, classic alternative, alternative rock, jam bands, and more).[7] There were also forty streams of news, sports, and entertainment.

XM's initial program grid included 101 channels, with 71 of those programming music (29 were news, talk, and information, and 1 was a free preview channel).[8] In late 2003 XM's lineup included six decades channels (1940s to 1990s), six country, eleven hits, two Christian music, twelve rock, eight "urban," seven jazz and blues, four dance, five Latin, five world, three classical, and two kids' channels. Within these larger genres or formats are subgenres or subformats. The idea was that XM and Sirius would *superserve* existing radio formats. For example, the six XM country channels included America (classic country), Nashville! (top country hits), X Country (progressive country, pronounced "cross country"), Hank's Place (traditional country), Bluegrass Junction (bluegrass), and The Village (Folk). XM's total channels increased in 2005 to 150, with 65 commercial-free music channels;[9] by 2008 it had increased its lineup to more than 170 channels.[10] More kilobits per second are required to transmit higher quality sound, and XM found ways to use its allocated bandwidth to increase its channel offerings. In 2008 XM boasted that its music programmers could draw from a centralized digital database of over 250,000 CDs and more than 2 million recordings.[11] That same year Sirius explained its use of hierarchical modulation to offer additional audio channels, "without noticeably affecting [its] broadcasts."[12] However, if one were to have listened to a channel like the Canadian CBC Radio 3,[13] they would have likely heard a decreased level of sound quality by comparison to a channel like Alt Nation on 36. Sound quality is not equal across the lineup and has to be negotiated as the lineup grows.

In 2023 SiriusXM maintained over 150 full-time channels, with over 140 of these available in Canada, though this did not include live sports channels. Its music channels included 25 pop (including decades and Latin music), 30 rock, 10 hip hop/R&B, 5 dance and electronic, 10 country, 3 Christian, 10 jazz/standards/classical, 2 family, 7 Canadian music, and a few other international music ones. Many of the genre-based channels are arranged in numerical order, but this isn't always the case. Of the 10 country channels, 8 are between channels 55 and 62. Further, there are music channels only available on the SXM app, like The Emo Project (emo) and Luna (Latin jazz). Music channels are organized around genres and time

Figure 16. XM channel lineup. *Source:* XM Satellite Radio, *2001 Annual Report*, 2002.

periods, like Classic Vinyl and Lithium (classic rock from 1965 to 1975 and 1990s alternative, respectively); by affiliation with a media company, like TikTok Radio and Pandora Now; by decades, such as 80s on 8; around a style of music, like The Blend or The Bridge (nice and easy pop and mellow classic rock, respectively); or around a band or artist. SiriusXM has artist-based channels for the following bands/artists: Pitbull, Elvis, The Beatles, Bob Marley, Bruce Springsteen, "Little Steven" Van Zandt, Jimmy Buffett, Pearl Jam, Grateful Dead, Phish, Dave Matthews Band, Tom Petty, U2, Ozzy Osbourne, Red Hot Chili Peppers, Drake, LL Cool J, Eminem, Diplo, Garth Brooks, Kenny Chesney, Willie Nelson, Dwight Yoakam, Frank Sinatra, and B. B. King.[14] More recently, channels have been added for Kelly Clarkson, John Mayer, and Carrie Underwood. On the SXM app, the following artists also have channels: Bon Jovi, Tom Petty (again), Marky Ramone, "Little Steven" (again), Steve Aoki, Miles Davis, and The Tragically Hip. These artists help to reach groups of listeners in disparate fan or taste communities, who are drawn to the exclusive access that subscription allows. Over time the number of channels available to subscribers has grown, but themes of distinction and expansiveness have been central tenets of the satellite radio business model from the very beginning.

MUSIC FROM THE SKY: DISTINCTION AND EXPANSIVENESS IN THE SATELLITE RADIO CHANNEL LINEUP

One profile of both satellite services in 1999 described their programming as one hundred "super-niche" channels, with music formats that could be "splintered to appeal to never-before-practical audience niches," thanks to the fact that the services reached a national audience instead of "the greatest common denominators" in smaller regions.[15] Along with new channels and new artists, proponents of satellite radio pointed to the ability of the services to bring back artists who were said to be forgotten on modern rock radio, like Patti Smith and the Replacements.[16] XM co-founder Lee Abrams emphasized that there was a "blank canvas to really rethink radio."[17] There was a desire to break away from the broadcast radio playbook and to think differently and with a fresh mind about what music programming could be. When I interviewed Abrams in June 2023, he said that XM delivered new formats that didn't exist on radio, or if they did, it was "part-time on some weird frequency." Abrams insisted that channels not simply be named "Rock 1," "Rock 2," "Country 1," and so forth, an idea that some individuals from the telecommunications industry had suggested. Instead, names needed to speak to the "mood, the energy of the channel."[18] What XM hoped to accomplish was to develop a nationwide audience for genres like reggae and blues. Popular radio formats were also expanded on satellite radio. XM used the oldies format to create multiple channels that were "so much different from what you hear on terrestrial radio." Some channels leaned into hit songs and others, like Deep Tracks, would venture deep into careers.[19] Abrams used the metaphor of the Mall of America to explain this approach: anchor stores (mainstream music) and specialty stores (jazz).[20] He added that the Deep Tracks channel would "go deep into careers," whereas on a top tracks channel one could "hear 'Bohemian Rhapsody' a lot."[21] One radio DJ that I spoke with in 2016 echoed this point, describing the channel The Loft as an "eclectic exploration of time through all sorts of fascinating [music]." Other channels might be more "pop focused" and have "tighter scheduling."[22] Recently, while listening to Deep Tracks, I heard Ellen Foley's "Thunder and Rain" (a cover of a Graham Parker song) followed by the Byrds' "Just a Season," the B-Side to "Chestnut Mare" and a much deeper track than a single like "Mr. Tambourine Man." Jumping to The Pulse, a contemporary pop hits channel, I heard Ed Sheeran's "Photograph," the sort of song one would hear on American hit radio. Over a random thirty-day period, the most played song on Deep Tracks had 9 plays ("Always with Me, Always with You" by Joe Satriani), while on The Pulse, the most played had 292 plays ("Love Like That" by Phillip Phillips).[23] This range reflects the sort of tension that exists in satellite radio because it reaches people broadly, coast to coast, but can also be hyper-focused on genres or categories.[24]

Because technological innovations shape our cultural experience and consumption habits,[25] there are ways that the technology of the satellite connected with beliefs about the virtues of expansive channel offerings. One profile of Sirius's appearance at the 2004 Las Vegas Consumer Electronics Show claimed that the company could experiment and take risks with new streams because there were no advertisers to "scare off" and because of investments in research and development.[26] The satellite and its connection to space and exploration articulates the expansive musical universe of satellite radio, particularly by comparison to its commercial radio competitors. These characteristics of the new services were neatly but aggressively packaged as a thirty-four-page, glossy insert contained in a special collector's edition of *Rolling Stone* from December 2003. The insert's front page proclaimed "Radio is Back! XM Liberates the Airwaves" alongside the headline: "Counting Crows Spread their Wings on XM Satellite Radio." Relatedly, the band's 1995 album is titled *Recovering the Satellites*. The next page unveils a two-page advertisement for Cadillac, a brand that was closely linked to satellite radio through XM's partnership with GM. Throughout the insert are five more pages of Cadillac ads as well as two pages of GM ads. A short "article" in the insert, titled "Introducing 'XM NATION,'" gives a brief history of music on the radio, highlighting when "powerful AM stations" pounded out a "diverse mix" of hits that was the soundtrack of America. FM then became the dominant radio band "by embracing the new music and culture, and by offering better sound quality." However, it continued, FM has grown "tired, narrow and repetitive," and "XM returns musical freedom to the airwaves."[27]

By emphasizing musical variety and sonically diverse radio, the satellite radio companies borrowed from freeform and college radio practices, like playing deep cuts and celebrating lesser-heard genres. But they repackaged this approach for affluent subscribers. Katherine Rye Jewell's *Live from the Underground* traces the history of college radio in America, the core of which is the period between the late 1970s to the early 2000s, when emergent genres like punk, new wave, and hip hop became central to college radio music programming. Jewell says that "college radio in the 1970s could be characterized roughly as alternative—leaving it to the DJs and programmers to determine what the term meant and how they would define the left of the dial."[28] These stations, Jewell adds, both defied and shaped the mainstream, which is a point that has also been brought up in writing on freeform radio. Michael J. Kramer explains that this type of radio was "a new hip capitalist style through which revolt against mass consumerism could be packaged and sold as a new market segment within consumerism itself."[29] Evidently there is a longer history of certain types of radio countering the dominance of commercial format radio, while also shaping the mainstream. Sirius and XM strategically employed a rhetoric of newness alongside these adventurous programming practices. What was novel in their new offerings, however, was bringing these sounds to national audiences on dedicated 24/7 music channels.

The classic rock format was one prominent example used to highlight existing radio deficiencies that Sirius or XM could remedy. One subheading in a 2001 issue of *Rolling Stone* read: "XM and Sirius think you hate your classic-rock station so much you'll pay to replace it." The article continued, "For decades, commercial radio has been shackled by inane DJs, tiny playlists and dull imaginations. That's about to change: Satellite radio is arriving this year, and it might not only improve your daily commute but also redefine a stagnant industry."[30] In 2005 Abrams, then chief programming officer of XM, was focusing on rock programming as an asset on the company's music roster. Abrams said, "The decline of rock on FM radio is of huge interest to us.... We play tapes of rock radio from the Fifties, Sixties and Seventies and think about how we can bring that back in 2005."[31] That same year, *Rolling Stone* writer Bill Werde said that radio stations were dumping the rock format as audiences declined and used the example of Y100 in Philadelphia as one major station that had shifted to the hip hop format early in 2005. Album-oriented rock (AOR) stations had listenership fall 70 percent since 1998.[32] Reports of the reduction in rock programming on commercial radio by the mid-2000s exemplified the ways that commercial broadcast radio was affected by trends in the record industry, namely the regrettable reproduction of formats and playlists across regions in America due to concentrated corporate power in the media industries. In 2001 more than half of all commercial radio stations used one of three formats (country, adult contemporary, or news/talk), and there were few American stations playing heavy metal, hip hop, or reggae full time.[33] In Detroit, seven of the city's most popular stations belonged to Clear Channel in 2004 and nationally in the US, the company owned approximately twelve hundred stations and about 70 percent of all promoted live events. As of 2004, Clear Channel's Top 40 stations shared more of the same songs than they had ten years earlier, and they relied heavily on the biggest hits; the top 5 songs heard on Z100 in 1994 were played between fifty and sixty times per week, and by 2004 they were played seventy-five to ninety times.[34]

Against this trajectory, Sirius and XM proclaimed the ability to offer multiple versions of a particular format, such as a large number of classic rock channels. They could also add or replace a channel on the channel guide quickly and regularly. This allowed Sirius and XM to respond to popularity and taste in the form of a channel devoted to a type of music, an artist, or even something like the emergence and popularity of podcasting. Sirius, for instance, began broadcasting a talk show by prominent podcaster Adam Curry in summer 2005.[35] In XM's early years after first rolling out its channel lineup, it responded to listener feedback and made programming changes to include channels dedicated to neo-soul, books and drama, old-time radio drama, folk music, and easy listening.[36] In summer 2005 XM dedicated a channel to reggaeton, called Algeria, responding to the genre's growing popularity.[37] It was a channel not originally offered by XM but that became essential to include. Sirius had also offered its own reggaeton show as of 2004, *El Rhumbon*, which aired on Sundays on the Wax channel.[38] The genre had been limited to specialized programs on Tropical broadcast radio stations but was a growing presence on many Spanish-language stations in the US. XM and Sirius, however, were able to deliver reggaeton music nationally and around the clock. In early 2004 Latin music in general was described as a growth area for Sirius, which had launched its third Latin music station in January of that year. Tropical was the newest, joining Mejicana and Universal Latino.[39] Executive VP of programming for XM Eric Logan, who was the former president of programming for Citadel Broadcasting, was credited for making effective programming moves.[40] At the same time, XM was dropping certain third-party channels, like MTV and VH1, a choice Logan said reflected the company becoming less dependent on third-party brands, indicating a sharper focus on exclusive, in-house programming.

The expanded universe of music programming on satellite radio tapped into a larger cultural context of indecision and variety in tastes that marked the turn of the century. Fred Goodman wrote in 2001 for *Rolling Stone*, "Call it democracy or just mediocrity, but in a year when the country almost couldn't pick a president, it was no surprise to find Americans just as divided on the direction of music." The year 2000's top-selling albums "were nearly an even split between rock, rap and pop performers."[41] These included NSYNC, Eminem, Dr. Dre, Nelly, Britney Spears, Backstreet Boys, The Beatles, Santana, and Creed. Thus, what satellite radio offered listeners was an expanded selection of channels that responded well to this level of indecision and variety. The overall channel lineup broke free from the organized *flow* of ad-supported commercial radio and aimed to entice and retain listeners in ways that reflect the cultural form of satellite radio.[42] The channel lineup constructs a world of controlled choice for which a subscriber can navigate. Channels change and adapt, with new ones being added and with "pop-up" channels appearing from time to time. Each channel has its own logic and strategy for keeping listeners tuned in, and some even cross-promote similar channels to help subscribers find others to tune into. In other words, channels can exist within one another's orbit. SiriusXM's current CEO, Jennifer C. Witz, says that the average subscriber has seven channels they regularly listen to.[43] Expansiveness allows for a feeling of comfort in the familiar as well as the feeling of being just a dial's turn away from another musical world.

The merged Sirius and XM channel lineup and music programming are still structured and shaped by a commercial logic that is bound to the record and radio industries. Notions of popularity and commercial and critical success, or value, shape the music one hears, and this feeling is largely subjective and tied to levels of taste or fandom. Channels with wider appeal typically fall lower on the lineup, while those with niche or limited appeal fall among the higher channel numbers. In spring 2007 the five most-listened-to channels on Sirius included two Howard Stern channels, The Highway (country), Sirius Hits 1, and Octane (hard rock). On XM, the five most-listened-to channels were The Blend (variety, pop), Flight 26 (now The Pulse, playing contemporary pop), Willie's Place (Willie Nelson's channel, now titled Willie's Roadhouse), Top Tracks, and Top 20 on 20.[44] All but the Howard Stern channels fall on the lower numbered end of the lineup. When XM and Sirius merged, programming overlaps and synergies were identified, and many of the more popular stations stuck around while others were removed or moved elsewhere in the lineup. Flight 26, for instance, was a very popular station, and the XM staff was kept on after the merger, even though the Sirius branding of The

Pulse became the channel's name. The channel also took a more contemporary pop tack, given its similarities to the 90s on 9 channel.

The satellite radio channel lineup offers an extensive array of music styles to appease a wide range of tastes and types of listeners, and as such it aims to sustain and grow subscriptions. Satellite radio, in order to convert listeners into subscribers, had to demonstrate value, and music plays a central function in accomplishing this. Perceived value drives subscriptions. It also makes one convinced of music's worth, particularly as it reaffirms taste and thoughts about why music is meaningful, or simply good. It is *perceived* because it is subjective but also shared, located in the packaging of music channels and communities of listeners.

SATELLITE OF LOVE: PERCEIVED VALUE ON THE CHANNEL LINEUP

Music connects with subscribers and reinforces their taste and self-identity. As such, it is central to the crafting of perceived value of subscription radio. Delineating and discussing value in popular music is often elusive in academic writing on popular music, in part due to the hierarchical and historical distinction between so-called high and low culture. Concepts of economic value or income-generating practices in popular music are also in flux. Scholars Hans-Joachim Bürkner and Bastian Lange argue that technological change and the internet have required an updating of concepts of value creation in popular music, and it is now proliferating in a "multitude of trial-and-error undertakings."[45] Popular music has been dismissed as merely serving commercial interests, seeking out the largest possible audience through formulaic structures and a dearth of innovativeness. But the reality is far more complex. Philosopher Alison Stone argues that popular music has aesthetic value; it centers the body and materiality, its repetition is central to its cultural form, and "simple songs can still be rich in meaning and interest."[46] Sociologist Motti Regev suggests that particular historical and musical eras, like the 1960s and rock music, have become valued or valorized as "masterpieces."[47] Popular music scholar Simon Frith locates pleasure, meaning, and value judgments in the way we talk about music, noting that to "be engaged with popular culture is to be discriminating."⁴⁸ Frith notes the absence of claims about the value of pop songs as objects, against assumptions that the value of such songs lie in what they can do for audiences.⁴⁹ Music has value in its role in shaping identities and our sense of self and in fostering sociability. Frith says, "The pleasures popular music offers us, the values it carries (and I include classical music as popular here), have to be related to the stories it tells about us in our genre identities."⁵⁰ Similarly, music industries scholar David Hesmondhalgh posits that music matters to so many people both because it feels emotionally linked to the private self and because it forms the basis of collective experiences.⁵¹ More recently, ethnomusicologist Timothy D. Taylor centers circulation in his thorough discussion of value in music, using Marx's idea that "money and commodities circulate everywhere" to talk about the circulation of commodities and how value can be tied to money, but also to "tokens beyond money," things like music fandom.⁵²

What is important to take away from these questions of value in popular music, at least for this chapter and its focus on music radio, is that music has value beyond its service as a commodity or as something that simply delivers profits to record labels or radio stations. Of course that's a large part of how value is understood and circulates, but I want to gesture to the construction of perceived value of music, a notion formed in subjectivities, emotions, and experiences as well as in distinction and value judgments. It helps to explain the way we perceive certain songs or artists to be of value, by way of becoming woven through identities and the stories we tell about ourselves. In other words, according to Frith, "we absorb songs into our own lives and rhythm into our own bodies."⁵³ This is why music matters and why we choose to listen to music on the radio. The relationship between music and our identity is used effectively by satellite radio to demonstrate value to listeners and subscribers. Frith, citing Theodor Adorno's "The Curves of the Needle," emphasizes this point in the example of the gramophone listener who only wants to hear themselves, and the artist merely offers "a substitute for the sounding image" of one's own person to be kept as a possession.⁵⁴ We find value in the music that speaks to us and forms our sense of self.

One way to communicate perceived value to listeners and subscribers is through exclusivity; namely, by acquiring musicians and celebrities to host or to brand channels after. Musicians affiliated with Sirius and XM in their early years included Grandmaster Flash, Jermaine Dupri, Del McCoury, Eminem, Quincy Jones, Bonnie Raitt, Tom Petty, and Snoop Dogg. Fittingly, George Clinton, an artist closely tied to the imagery and imaginary of outer space, had "connected with the mothership—that is, satellite radio broadcaster Sirius" to host on the Express Channel.[55] Some artists, like Jimmy Buffett, brought a clearly defined and exceptionally passionate fan base to satellite radio. Buffett's Radio Margaritaville operated as an online venture for some years before it found a place on Sirius Satellite Radio. Steve Blatter, the senior vice president of music programming at Sirius in 2005, said that Radio Margaritaville would be the first time Sirius subscribers could get the channel in their cars or their boats,[56] a listening space important to the leisure-by-the-water-loving Buffett fan base ("Parrotheads," who enjoy listening to Radio Margaritaville while eating a cheeseburger in paradise).

Artists and celebrities are present in the program grid in different ways. A number of channels are based entirely on a band or artist. Others simply host a show that runs for an hour or two, lending their celebrity capital to a particular channel for a portion of the week or month. The artist-based channels craft a 24/7 music experience that encompasses an artist's music, persona, and lifestyle. These channels are reserved for those artists thought to have established lifestyle elements to accompany their music or those with an incredibly dedicated fan base. In reference to Eminem's Shade 45 channel, then president of entertainment and sports Scott Greenstein claimed, "This is the most direct way an artist can reach fans. It takes time for an artist to put out a new record, [but] this is a way to connect 24 hours a day."[57] The platinum-selling Eminem played into the uncensored nature of Sirius's programming after his song "The Real Slim Shady" (2000) was briefly deemed indecent by the Federal Communications Commission (FCC).[58] Artists on air lend an aura of exclusivity to satellite radio. Although artist hosts are commonly heard across the broader radio landscape, the high level of resources enjoyed by the satellite radio industry has meant that the artists who host shows are of significant notoriety and fame. In 2006 Tom Petty hosted a freeform-style radio program on XM called *Buried Treasure*, which eventually became its own channel. The first single and title track from Petty's 2002 album, "The Last DJ," was used as an opportunity to bring Petty to XM. This song was appropriate because of its focus on "the state of terrestrial radio at the time." The song generated controversy in the radio industry, with some stations feeling that it was a "slap in the face." Many stations decided not to play it. Petty said, "I don't give a flying fuck about any of it. I've tuned out. But I was elated when my song was banned.... I remember when the radio meant something."[59] Abrams said that Petty was ideal for XM: "He's the kind of artist we like on XM because he's intelligent, and he's also musically diverse, and he's not just going to play a bunch of classic-rock hits. It's really music from his soul."[60] The program served as an entry point into exclusive Petty-related material, such as an "eight-hour historical retrospective of Tom Petty and the Heartbreakers."[61]

Satellite radio also asserted the value of music by commercial means, maintaining long-standing relationships between the record and radio industries and propelling the culture industries into the new millennium. As one example, Jimmy Iovine, then chair of Interscope/Geffen/A&M, was executive producer of the Shade 45 channel and consulted Sirius on new programming opportunities. In 2005 he said that the "opportunity that this relationship creates for an individual artist, a record label or, for that matter, an industry, customizing distribution and exposure, is unparalleled in the history of the music business."[62] The radio and record industries are historically intertwined in a system of hit-making, promotion, and the forming and targeting of audiences. Popular music scholar Keith Negus explains that the "radio networks of North America and Britain provide one of the most important promotional outlets for recorded music, setting programming agendas at radio stations and influencing the talent acquisition policies of record labels throughout the world."[63] Between 1920 and 1950 a symbiotic relationship between the radio and music industries developed, with music becoming the key element of radio programming and radio programming serving to promote records. Radio became integral in providing a livelihood for musicians and composers.[64] In the decade before the beginning of World War II, the major phonograph companies of Victor Talking Machine and Columbia were dependent branch divisions of the two large American broadcasting networks.[65] Until the late 1940s RCA and CBS "determined which kind of music could be heard on their radio stations, controlled the right to exploit their music through their publishing houses, and defined who was and was not allowed to have access to their distribution networks."[66] Over the years, the radio industry has used genre definitions to build formats and locate audiences, a process that aligns, at times, with the categories used by the record industry. The radio format imagines a particular audience for its music and may embody the conventions of a particular genre, in the case of rock or country, or may encompass a number of genres, like Top 40. In any case, radio programmers guide the actions and activities of record companies.[67] From the 1950s onward, the medium was central in creating hit songs.[68]

As Sirius and XM established themselves as new audio services, their centrality to the record industry became apparent. In 2005 it was said that the record industry had started recognizing "the enormous power and reach" of satellite channels and that the flexibility that comes with satellite technology and radio programming gave "the labels and the artists the unparalleled access and the unparalleled opportunity to do things that have never been done before."[69] Years before, some XM employees referred to the company as a "Secret MTV" in terms of its groundbreaking nature, which "reshaped the musical landscape as it helped make a number of careers."[70] These optimistic sentiments followed a brief period of uncertainty on the part of labels with respect to what satellite broadcasters had in store for the industry. Part of this was the question of subscriber numbers and what the uptake of the satellite services would be.[71]

Record label and satellite radio synergies are particularly apparent in SiriusXM's "pop-up" channels. These channels temporarily replace a full-time station for a period of time, although some have been added to the growing online-only channel roster. On July 20, 2020, a number of these channels replaced regular programming for a limited time. These included The Queen Channel, The Michael Jackson Channel, The Beastie Boys Channel, and The Prince Channel. What these artists have in common is that they are all major legacy acts with extensive catalogs who are deceased, or bands with key members who are. Their catalogs exist as assets to be mined and used to continually generate revenue for rightsholders. They are also artists about whom major motion picture biopics or documentaries had recently been made or about whom such films were in the works. Queen was enjoying massive back catalog success in the wake of

Bohemian Rhapsody, with Universal Music Group reaping much of this benefit. Queen's members were huge streaming stars after the film, becoming the most streamed rock band of the twentieth century and "Bohemian Rhapsody" the world's most-streamed classic rock song of all time.[72] The film's producer also made a deal with Michael Jackson's estate in 2019 for an authorized biopic.[73] The Beastie Boys channel followed with an Apple TV documentary directed by Spike Jonze, and a movie about Prince was in the works as of December 2018 at Universal Pictures, which had picked up the rights from Universal Music Publishing to a number of his songs.[74]

Within this duality of value, where on the one hand perceived value accounts for subjective and affective feelings of music listeners, and on the other hand is the monetary value that subscription radio generates for the music industries, it's worth taking a closer look at the role of artists and their related channels across the lineup. Channels, whether pop-up or full time, serve to promote living and working artists during and between major album releases and tours. The U2 radio station, U2 X-RADIO, was announced in 2020 during The Joshua Tree Tour while the band was on stage in Tokyo, Japan. The announcement was full of synergies, with an appearance from SiriusXM's Howard Stern via video link, which was also streamed on the U2 Facebook page.[75] On the launch of the new channel, guitarist The Edge said that the channel would be distinct from streaming, "more curated" and with a "personal connection with the artist." The Edge was given his own show, *Close to The Edge*, which involves "a series of deep-dive conversations about the creative process."[76] The satellite radio companies also make use of their studio spaces for the purpose of record promotion. Major artists like Duran Duran and Elvis Costello used the Sirius New York headquarters in September 2004 to promote their respective albums.[77] Moreover, these visits yielded exclusive in-studio performances that appropriate music channels could air regularly and use to continually promote the album and take advantage of in-house, exclusive music programming.

Over the years the story of music programming on Sirius, XM, and SiriusXM has been one of industry partnerships, fan service, and finding growing markets that aren't reflected on commercial broadcast radio. Much of this is a public relations style narrative meant to continually reinforce and exert the service's perceived value against the potential shortcomings of satellite radio's competitors. But over the course of satellite radio's existence, the relationship between the radio industry and record industry has been reworked to put satellite radio at the center of this long-standing partnership.

FORMATS BEYOND FORMATS, NOSTALGIA, AND MUSIC HERITAGE ON THE RADIO

In this chapter so far I have discussed the organization of music programming on satellite radio, particularly in its early years. Artist-based channels are one way that perceived value is crafted for subscribers, drawing on taste and subjectivity to connect artist catalogs to listeners. Big-name artists also propel the cultural industries by keeping catalogs relevant, or valuable, as they circulate across the lineup. The expansiveness of the lineup involves further strategies for connecting with listeners. Perceived value is also crafted by lending music an aura of authenticity and exclusivity. In some cases, these practices or techniques amplify, or modify, the sort of programming one might hear on commercial, public, or community broadcasting. In others, they make more extensive use of the resources that the satellite radio companies enjoy. Although a virtue of the satellite channel lineup is its expanded choice compared to what commercial radio offers, it still aims to funnel listeners to their favorite channels and artists. What ascribes value to satellite radio programming, in the minds and ears of listeners and subscribers, is the capacity to tap into music's personal and intimate connections to one's sense of self, or the "organization of self," both in the present and in the past.[78] Four of these on-air techniques are playing superserved formats, celebrating neglected formats, crafting audio ambience channels, and broadcasting on-site experiences. By drawing on nostalgia, music heritage, fan cultures, and sounds that have been neglected by commercial broadcast radio, the satellite radio industry uses music and its emotional and meaningful qualities to reach and secure a base of subscriber-listeners across taste cultures, ages, and demographics and within vast streams of music. This serves to maintain and increase subscribers by, ideally, convincing subscribers of the value of subscription radio.

As I have indicated, a key aspect of the marketing and promotion for the new satellite radio services was its stated superiority over commercial radio regarding music programming. Radio formats define commercial radio stations by delineating what music and which artists are included in music programming, and they work to categorize listeners into demographics or market categories that can then be sold to advertisers.[79] Formats determine not only the type of music heard on a given station but also the other "ingredients" in a given program hour, such as advertisements, radio voices, and so forth.[80] Some of the limitations of commercial radio formats that satellite radio was said to overcome are a high level of song repetition, music that caters to advertisers, and the fact that many styles of music did not fit within existing formats. Professor of English Jeffrey Roessner has written that satellite radio's expanded playlists challenge both genre and format, going beyond "subgenres in favor of a proliferation of micro-genres."[81] Roessner adds that "it may be more helpful to think of many of the satellite channels as representing micro-genres that have a rhizomatic relationship to broader, and deceptively stable, genre categories such as jazz, country, or rock."[82] This means expanding the programming choices for a given radio format. The satellite radio companies describe this as superserving music formats.

For example, oldies music is often aggregated on a single format for broadcast radio. XM segmented this category by programming multiple channels devoted to decades from the 1940s to the 1990s. In 2005 it was said that the decades music channels were programmed from the top charting hits from each decade, with a regular playlist of more than three thousand regularly played songs and another two thousand that were played occasionally.[83] While many of the radio formats available on satellite radio extend from already existing radio formats, what is unique about the satellite channels is the subformats or subgenres that orbit the larger, more familiar formats or genres. Rock, for example, is segmented into channels like 1st Wave (classic alternative), SiriusXMU (indie and beyond), and Hair Nation (80s hair metal and glam). There are also artist-based channels that fall under rock, like E Street Radio (Bruce Springsteen), The Beatles Channel, and the Grateful Dead Channel. In jazz programming, Sirius started with six jazz channels and XM with seven. Channels showcase subgenres, like classic swing or fusion.

Mike Peters, who was the genre manager for Sirius in 2003, said that each "stream satisfies the tastes of a large national group of jazz fans. Unlike conventional radio, which has specialty shows devoted to a particular type of jazz, a [satellite] listener can tune into a station that plays only what they want to hear, 24 hours a day."[84]

Superserving formats connects with the notion that satellite radio is unshackled from restrictive commercial radio formats, or, at least with the idea that formats are highly limited. Eric Weisbard complicates this trajectory and its related hierarchization of culture, in which formats and their commercial imperatives are thought to be of little value. Weisbard argues that formats are more diverse than they are often given credit for and enable rival mainstreams to thrive in pop culture's center. Using Dolly Parton as an example, he writes that women were central to country as a radio format despite its being marginalized as a genre.[85] However, as Jada Watson outlines, women country artists have found success in crossover airplay but have had little impact on the country format itself.[86] And there are certainly limits to what one station can accomplish with respect to music programming with the ad-driven structure of commercial broadcasting. Without a target demographic in a given region, or market, some stations or styles of music do not make economic sense to the commercial model. This means that a major city like New York City or Los Angeles could be without a country format station for a period of time. In August 2006 Lee Abrams wrote, "No Country radio in LA anymore. There are 2,782 Pop stations though. No Metal station there either."[87] Radio format categories influence the type of artists played and presented;[88] thus, a more expansive channel lineup enables a wider variety of artists to be heard.

By subdividing and superserving formats, music that is popular and sells well outside of radio, either through the sale of recorded music or in ticket sales for live performances, could find a spot on the Sirius or XM channel lineup. Over the years of Sirius's and XM's operation, hip hop has been a massively popular genre. It also has often been mixed in with genres like soul or R&B to form the unfortunately named "urban" radio format. In some cases, hip hop artists have faced difficulty in finding airplay on Top 40 or hit format stations, despite success on streaming platforms. As music radio scholar Amy Coddington writes, many hit radio stations have a history of playing hip hop hits, while others have been "holdouts against rap's radio ascendance."[89] Between 2012 and 2018 only one non-white rapper, Drake, had successfully crossed over from the "urban" radio format in the US to top the Pop Songs chart.[90] The sort of freedom from indecency regulations that helped bring Howard Stern to Sirius also worked well for hip hop lyrics, which were prone to censorship or avoidance by more conservative radio companies or regions. XM programmer Leo G said that its Raw channel emulated hip hop lifestyle with a "mix-tape mentality," one that Sirius followed, having DJs around the country submit mixes for airplay.[91] The streaming era has demonstrated that there is a massive audience for hip hop, despite commercial radio not always embracing the genre over the past few decades.[92] Sirius and XM anticipated the streaming success of hip hop in their early circulation of hip hop music.

XM also explained that it sought out music formats with a strong demand but that were relegated to AM stations due to lack of availability on the FM band. These formats included classic country, music of the 1940s, and gospel. XM has dedicated channels for these formats.[93] In 2006 XM said that it offered music "with significant popularity, as measured by recorded music sales and concert revenues, which [is] unavailable in many traditional AM/FM radio markets" such as classical, blues, heavy metal, electronic dance, disco, and jazz.[94] One 1998 study shared by the RIAA indicated that recorded music sales of niche music formats like classical, jazz, and Broadway soundtracks comprised up to 21 percent of total recorded music sales in 1998.[95] Thus there is a clear commercial advantage to this approach as well.

Sirius and XM also invested heavily in the country music format. The satellite radio companies framed their focus on country music as an act to preserve and revitalize the format, one they felt was in dire straits in the mid-2000s. New York City, for instance, lost its last dedicated country music station in 2002.[96] In 1995 *Billboard* claimed that country radio ignored the music of independent label acts, finding it difficult to compete against major label artists and their close relationship with the radio industry.[97] The 2004 Country Radio Seminar in Las Vegas focused extensively on "ways the country format can beat the doldrums" and overcome a struggling industry. The head of country programming for Sirius, Scott Lindy, said that the company was "not going to let this format die." Lindy "encouraged the industry to take more chances and be more experimental."[98] Bluegrass was one subgenre of country that Sirius and XM programmed on their own channels. Lindy explained that broadcast radio was lacking in full-time bluegrass stations, with the genre only being heard for a few hours here and there on certain FM and AM stations.[99] The International Bluegrass Music Association said in 2005 that bluegrass was heard on around eight hundred stations that are programming at least six hours a week, split equally between commercial and noncommercial stations.[100] Bluegrass has numerous fans across the nation, "millions," according to Lindy, but was without a consistent presence on commercial radio across the US. One of satellite radio's earliest musician-hosts was "Little Steven" Van Zandt from Bruce Springsteen's E Street band and HBO's *The Sopranos*. Van Zandt is the celebrity behind the Underground Garage channel and became involved with Sirius as a creative adviser in January 2004. He was also a key figure in developing the Outlaw Country channel in 2004. The purpose of the channel was to feature music that was "falling between the cracks." Van Zandt had a long list of songs that he wanted to play on the channel, which Sirius staffers were having trouble locating. But DJ Jeremy Tepper, a music journalist and record collector, ended up finding them in his own collection. Outlaw was envisioned as more of an attitude than a genre or format and thus was said to defy "the boundaries of traditional radio formats," being purposefully "wide in style and era," connecting the dots between "Texas swing, rockabilly, country that twangs, country rock, alternative country and three generations of Hank Williamses." The Outlaw station draws from a music library that stretches back to the 1930s but also programs new artists and songs.[101]

As with country, the satellite radio companies were keen on electronic music in their early years. Electronic music in its broad form, that of the widely popular and largely UK-based electronica of the late 1990s or the more recent umbrella term of EDM, has been massively influential on popular culture, but it's a sound notably absent on commercial radio. In 1999 a report on *Billboard*'s Dance Music Summit explained that there has been a persistent lack of support for dance artists by both radio and MTV. Thus, the internet was becoming a key resource for dance artists to generate exposure in America.[102] There are certainly shows that have championed electronic music, namely on public and campus/college radio, but the duration of its tracks, its connection to the dance floor, and its lack of vocals make it out of step with the sort of starmaking and lyrics-based focus of commercial broadcast radio.[103] Sirius and XM recognized the strong market for electronic music beginning in the early days of the companies. Satellite radio was cited as a new tool with great potential to showcase dance in America. LA-based radio programmer and club DJ Swedish Egil was on board early with Sirius, then CD Radio, to program its dance channel, and he said it would "offer music not likely to be embraced by traditional commercial broadcasters."[104] Before XM's launch, dance music program director for the company Blake Lawrence unveiled plans to have at least four channels devoted to dance music, with one focused on heritage dance. The others would draw from "underground house, commercial sounds, and electronic."[105] Lawrence pointed to a divide between the music being played on XM and the dance music record industry in 2003, explaining that XM often played dance remixes that listeners could not locate to purchase. He said that XM radio hosts suggested stores and online retailers where listeners hopefully could find the songs, but that "sometimes the particular mix they're looking for simply doesn't end up at retail."[106] As with hip hop and country, EDM remains prominent across the music channels on SiriusXM today.

For subscribers, hearing music that is not available on most broadcast radio stations is one way to justify the cost of subscription access. Another is to find affective connections with one's favorite music, especially music from one's past. Media, as they pass us by or persist and adapt over the years, are intertwined with feelings of nostalgia, as they are tools that mediate our experience of music in ways connected to the passing of time. Media scholar Kyle Barnett writes on the paradox of old-time music on the radio, an antimodern genre that critiqued ideas of progress in the 1920s but found acceptance on the new technological medium of radio.[107] A similar paradox exists on the satellite radio channel lineup, where numerous stations are anchored in a moment in the past and radiate its cultural and musical sensibilities and minutiae, yet they serve to promote a new service that claims to be the radio of the future. The themes of space and alien life that come with the satellite and its place in orbit also promote a feeling of nostalgia. Ken McLeod says that a "musical end-of-the millennium angst" brings forth a fascination with alien themes informed

Figure 17. Ad for Electronic Dance channel on XM Radio. *Source:* XM Satellite Radio, *2001 Annual Report*, 2002.

by "a postmodern erosion of historical narrative of progress." This is found in the "temporal, melodic and timbral dislocations that typify late-1990s alternative musical style and in a turn to nostalgic reifications of previous rock eras."[108] Satellite radio, with its simultaneous use of futuristic and outer space themes and of ideological notions of legacy and authenticity and cultural value tied to music across history, fits within this particular late 1990s musical-cultural context and brings it further into the twenty-first century.

One aspect of the Sirius and XM channel lineups that stands out as being tailored to subscriber subjectivities is the decades music channels and those that are rooted firmly in a musical-historical moment for which to craft an overall ambience that conveys the sounds and feelings of a past era. Music journalist Simon Reynolds, in a critical take on the present's inability to imagine or enact a future, due largely to the prominence of nostalgia in popular culture, claims that the future will "just consist of *more of the exact same*."[109] In this way, the satellite radio channel lineup makes sense as simultaneously able to hold a wide range of musical eras together. Music serves as a gateway to past eras and is particularly effective at tapping into nostalgia. It has "an unparalleled capacity to distil the atmosphere of a historical era."[110] In expansive box sets and remastered and re-released albums, the past is glorified and given value. These cultural artifacts are complete with photos, interviews, and visual and textual materials that serve as an "in-depth history book."[111] Packaging music according to decades is also a reliable commercial framework for the cultural industries to circulate and sell back catalogs of music and for journalists and other media workers to write about and make sense of history.[112] Channels rooted in the past, those that the satellite radio companies have referred to as *audio ambience* channels, tap into this association between past moments and their musical touchstones to take on the sound of an older radio station, going beyond the act of just playing songs from across the decades.

Audio ambience channels use audio signatures and sounds to create a sonic atmosphere.[113] Examples of these sonic characteristics include a reference to a five-disc CD changer on the 90s on 9 channel to indicate an upcoming "random" selection of songs, as well as having 1980s MTV VJs host on the 80s on 8 channel. In the latter case, the voice and vocal delivery of the radio DJ suits the era from which the music is chosen.

The Classic Vinyl channel communicates a format-based time frame, as does the cassette-era classic rock channel, Classic Rewind. These channels tap into a larger feeling of the era that first programmed the music, and channel promos speak to the listener with a high level of insiderness. One promo that aired on the 1990s grunge channel Lithium played Mazzy Star's "Fade into You" and then asked the listener to recall a fling from the 1990s, one in which they couldn't recall the partner but certainly could remember their song.[114] It ended by describing the channel as "alternative and grunge engulfed in nostalgia." In most cases, the efforts put into crafting this audio ambience are minimal and reliant on on-air personalities and brief references to music formats and things that took place in a certain year or decade. In other cases, channels will air radio programs from the past. For example, *Casey Kasem's American Top 40* has aired as much as three times a week on the 70s on 7 channel. In another example, the host of *Hipster Runoff Blog Radio* on SiriusXMU relied on an over-the-top ironic and detached vocal style to talk about millennial hipster culture in between sets of chillwave music. One reddit user described these shows as "the value [added]" that they pay for in a subscription.[115]

According to Abrams, the 60s channel aimed to mimic a "Top 40 station in about 1965," and it "was loaded with talent, loaded with personalities." He said, "Our 60s channel, for example, really screwed you up. You listen to it and go, 'What, am I back in 1965?' Everything from reverb to the authentic playlist, the authentic jock presentations.... It was really compelling." The term *audio animators* was new for production people, Abrams said. It suited the sound because "we're animating the sound," working to move away from the standard clichéd sound of radio. On the traditional country channel, Willie's Place, the aim was to craft a sound of authenticity, according to Abrams: "Smell the speakers and you'll smell chewing tobacco, stale beer and worn out 45's from an old Wurlitzer. Modeled after the way an AM Country station in Lubbock probably sounded circa 1955." On oldies channels, XM used "memory cluster" music sequencing, with songs clustered in "three song sets from a given season, so memories will BURN IN and are not simply fleeting moments as the typical oldies station navigates from era to era."[116] This use of older sounds and technologies to tap into nostalgia and into music's unique ability to bring a listener back to the past, or at least the past as remembered by a particular listener, is a sort of mediated nostalgia, "nostalgia generated through the record player, the photograph, or the iPod," or a notion of *sound souvenirs*, the "cherished tapes and albums we keep in our attics."[117] The emotions and memories we link with music allow us to reflect on and revisit the past upon hearing the right song at the right time. Music, and the way we remember it, is shaped by technologies and formats of the day. Material formats have a tactile sensation, one that certain channels attempt to bring to life, evading the lack of potential emotive contexts that some argue comes with the turn to digital sound files or streaming music.[118]

The concept of simulacrum contextualizes the audio ambience channels and their referencing of the past through constructing particular sound worlds that aim to communicate a *feeling* of the cultural and technological era in which music was first heard. Jean Baudrillard wrote that through the collapse of "the real" came the capacity for nostalgia to assume its full meaning.[119] Thus begins the era of hyperreality. In reality's void is the simulacrum, where there is only simulation, a perceived reality, copies that no longer have an original. In Gilles Deleuze's formulation of the simulacrum, he writes that it still produces an *effect* of resemblance but one that is wholly external and constructed around a disparity: "That is why we can no longer even define it with regard to the model at work in copies."[120] Across the satellite radio channel lineup, listeners seek points of reference, familiar entryways into the past, to historical eras or moments that are long gone. There is a comforting familiarity in hearing touchstones, familiar artists and cultural stewards who usher in a reflection of a time gone by or a time once lived. Audio ambience channels aim to represent the feelings of the real, or of a historical moment long gone.

Once when listening to the 50s Gold channel, I heard a long comic monologue, "It's in the Book" by Johnny Standley. It's rare to hear a monologue like this on a decades channel, but in this case it served to bring back a voice from over seventy years ago, collapsing this chasm of time. It provided that audio ambience. And certain songs come to stand in for a decade and communicate its essence, no matter how deceiving this may be. Raymond Williams uses the concept of selective tradition to explain the ways that historical moments are routinely defined by particular cultural artifacts at the expense of everything else from the lived realities of past eras.[121] The 1960s has been reduced to signs, symbols, and sounds that communicate the decade. A channel like 60s on 6 (now 60s Gold) or Classic Vinyl, through select songs and sonic cues, constructs an idea of the 1960s that is convincing to listeners who wish to relive or experience a select array of meaningful cultural aspects from this era. Audio ambience channels recreate radio stations that do not exist or never actually existed. When navigating the service's online app in 2024 this became even more apparent, as eras were reduced to playlist icons that featured images like an old radio receiver for the 50s Gold channel.

The era of the satellite, or the space age of the 1950s and beyond, also begins Fredric Jameson's periodization of postmodernism, "after the wartime shortages of consumer goods and spare parts had been made up, and new products and new technologies (not least those of the media) could be pioneered."[122] For Jameson, the postmodern culture of this period is the "superstructural expression of a whole new wave of American military and economic domination throughout the world," within which aesthetic production "has become integrated into commodity production generally," with "the frantic economic urgency of producing fresh waves of ever more novel-seeming goods."[123] In this formation, nostalgia thrives as "the producers of culture have nowhere to turn but to the past: the imitation of dead styles, speech through all the masks and voices stored up in the imaginary museum of a now global culture."[124] The satellite and its articulated range of music programming can be understood through Jameson's critique, as they extend from military power, they represent American global power in the form of its monopoly status (particularly in the SiriusXM era), satellites require wealth and political power to develop and launch, and music programming connects with a listener's memory and fondness for nostalgia through recreating the appropriate ambience for grounding musical selections from the past. As Mark Fisher writes, "The only future that capital can reliably deliver is technological—we count historical time . . . in technological upgrades, watching the same old things on higher definition screens."[125] Or we hear the same things on higher definition radio. The notion of progress that Sirius and XM promoted throughout their advertisements and press releases would be found in the higher quality of the radio service's sound and in the expanded bandwidth of the channel lineup. With Sirius and XM responsible to shareholders, innovation and experimentation would be tempered by a market logic that ensured subscriber growth by not straying too far from already existing forms and logics of value in popular music culture.

One other place to hear how perceived value is crafted on the channel lineup, namely by association with some sense of an "authentic" musical experience, is via on-site radio programming, or the telegraphing of culturally rich music heritage sites. With satellite radio, a particular orientation of looking, thinking, and listening "up" is assumed, given the orbiting device in outer space that is the key infrastructure for delivering radio to listeners. But on-the-ground infrastructure is just as important in terms of bringing a certain type of music programming to life. Major investments in studios and partnerships with institutions that have already established a high level of value and prestige in their own right further enable satellite radio to craft perceived value through programming. On a channel like SiriusXMU, live *XMU Sessions* profile and promote an artist's new music, bringing a performance from the SiriusXM studios to the airwaves. These sessions typically result in an exclusive live version of a song that then receives airplay on the channel. XM routinely proclaimed its ongoing investment in live performances, interviews, and exclusive music programming through offerings like the *Artist Confidential* series, which showcased live performances and interviews with artists. By 2005 artists such as Paul McCartney, Coldplay, and Bonnie Raitt had performed in XM's studios in front of a live audience, and the company had hosted more than one thousand live performances.[126] The listener, then, is brought into the studio space. Only the subscriber has access.

One notable partnership between XM and the Country Music Hall of Fame in Nashville was announced in the summer of 2000. It involved XM broadcasting a live, five-hour show daily from a new digital studio in the Hall of Fame's museum. Resulting from the partnership was XM's access to the Hall of Fame archives, which include over two hundred thousand recorded discs and hundreds of audiotapes. These recordings were used for programs like *The Country Music Hall of Fame Hour*, which featured profiles of country music legends and rare, archival recordings.[127] Through such strategic partnerships, music programming reflects efforts at linking popular music with a notion of heritage and cultural value;[128] it also reflects the travel and music tourism that has come with the growing significance of music in regional cultural economies.[129] The museum or the hall of fame absorbs particular selections of music, saving it to be remembered, to not be discarded, forgotten, or thrown away.[130]

Along with the XM and Country Music Hall of Fame partnership, both Sirius and XM established footholds in Nashville, an important move given the city's strong music industry activity. In 2004 both Sirius and XM announced plans to increase their presence in the city. Sirius hired Scott Lindy from WPOC in Baltimore and relocated him to Nashville as director of country programming. Sirius also hired a Nashville music industry veteran, Charlie Monk, to develop talent and programming for the country streams and to act as a liaison to the Nashville music community.[131] XM hired former WMZQ Washington, D.C., music director Jon Anthony as its music director and on air personality for Highway 16, one of its country channels, who also relocated to Nashville. Sirius took on office space in Nashville and built two studios there, while XM had already built its Hall of Fame studio. Lindy said, "If we're serious about country and we're not in Nashville, who's going to take us seriously?"[132] XM continued to expand its country music infrastructure in 2006 with the addition of Willie Nelson as a "proprietor" of the new Willie's Place classic country channel. The channel began broadcasting from new studios at Nelson's BioDiesel Truck Stop in Carl's Corner, Texas, in 2007.[133] The new facility included two restaurants, a saloon, and a 750-seat performance hall, along with the XM studio.[134] XM also connected with Jazz at the Lincoln Center as its exclusive satellite radio partner.[135] Evidently, its rental and ownership of studios and spaces, its brick-and-mortar footprint, served to reinforce and ascribe value to its music programming.

MUSIC LIFE CYCLES

In "Atlantic City" from Bruce Springsteen's haunting 1982 album *Nebraska*, Springsteen sings, "Everything dies, baby, that's a fact / But maybe everything that dies someday comes back." What much of the satellite radio channel lineup accomplishes, in addition to those channels that propel and promote recent releases that set ablaze the public relations, promotion, and marketing machine, is the ability to keep songs alive and to bring the (seemingly) dead back to life. The cycles of music genres and music formats can be said to reflect a "trope of death" throughout their discursive histories.[136] As Fisher writes, "Cultures have vibrancy, piquancy only for a while. Lyric poetry, the novel, opera, jazz had their time; there is no question of these cultures *dying*, they survive, but with their will-to-power diminished, their capacity to *define* a time lost. No longer historic or existential, they become historical and aesthetic—lifestyle options not ways of life."[137] These historical and aesthetic options are not equal, however, and institutions and intermediaries play a role in cycling songs in and out with respect to popularity, profitability, and relevance. Cultural studies scholar David R. Shumway refers to this as commodified nostalgia, the "revival by the culture industry of certain fashions and styles of a particular past era."[138] The expansive satellite radio channel lineup is uniquely well-positioned to do this.

Nostalgia also helps to filter and shape the playlists preferred by these channels, as they prove to be appealing to listeners and help retain subscribers. Studies have shown that taste preferences for popular music often form during late adolescence and early adulthood, and these preferences often prevail for much of one's life.[139] Abrams affirmed this notion and told me that "the music people like between 16 and 20 is what you like for life." XM created "formats that focus on where people were when they were 16 to 20, depending on your age." There's a 40s channel for people who "were 16 to 20 during the big band era."[140] The channel lineup can host a range of music eras that speak to subscribers across generations. Subscribers can access formative musical and cultural moments on a daily basis.

My interest here isn't in establishing a criterion for what is valuable, or *good*, or in advancing a new concept of value for music, but rather to acknowledge the fact that across the vast history and range of recorded music and in the ever-expanding channel lineup of a service like SiriusXM, some songs are in and some are out, or alive and dead. Some artists are monolithic in their likeness, becoming crystallized as a radio channel, others are one of millions of entries in a digital database, and others are not in the database at all. Selections are made. Institutions establish criteria, programmers make choices, all shaped by a combination of industry trends and expectations, ideas of marketability, legacy and myths, taste, gut feelings, emotions, and memories. While it is fair to conceptualize the organization of channels on Sirius and XM as expansive, by comparison to commercial broadcast radio particularly, there still exists a level of repetition and predictability on each channel. If you routinely listen to the same channel, you will hear the same songs regularly, on some channels more than others. This mirrors the sort of record store that uses space to highlight and showcase select albums to entice "high-value" or prestige customers, adding value to the cultural products, as opposed to the one that might pile albums from floor to ceiling with no discernible plan or purpose.[141]

The moment at which satellite radio launched is important for thinking through the centrality of cycles, or orbits, of music and culture and their place in shaping the organization of music programming on Sirius and XM. A parallel example from this era is the indie resurgence in North America, articulated by new industry relationships as well as a decline in power and gatekeeping by the radio industry. A song like Metric's "Dead Disco" can be placed within an indie dance/rock sound that was prominent across the 2000s, one that was especially generous to indie bands in Canada, who reached a national and international stage with immense success during this era. Along with Metric, bands and artists like Stars, Broken Social Scene, and Feist all flirted with mainstream success, some becoming fully engulfed within it, a particularly lucrative moment for independent bands with small label representation. "Dead Disco" (2003), certified Gold in Canada in late 2005, kicks off with prominent dance-forward hi-hats, reminiscent of disco but also of dance-inspired new wave and post-punk of the early 1980s. The chorus hits with an ironic, detached vocal line by Emily Haines, who laments: "Dead disco / Dead funk / Dead rock and roll / Remodel/ Everything's been done / La la la la la la la la la la la"; the "las" become increasingly disinterested and distant as they go. While the lyrics of the chorus express a 2000s hipster too-coolism and frustration with the recycling or "remodeling" of styles of music, "Dead Disco" was not immune to this either. The dance-forward sounds of the track, coming together with energetic, high-tempo drums and bass, were a remodeling of UK new wave and punk, just as bands like the Strokes and White Stripes, of the same era, were drawing on New York and Detroit garage and proto-punk. The music blog *Drowned in Sound* calls the album "fresher and more intelligent than what might be construed as yet more retro-revivalism."[142] *Pitchfork* says, "Metric melds together the usual suspects (The Cure, XTC, The Velvet Underground, New Order) for a new wave-tinged exploration of off-kilter indie rock."[143] One 2021 review from *Midrange Vancouver* places the song in the post-9/11 millennial era, writing that there is a "searing anger burned into James Shaw's guitar during the intro," helping to rail "against the cultural dead ends of late stage consumer capitalism."[144]

The recycling of sounds, paired with the ongoing and persistent release of new music and all the processes of filtering songs for radio play and hit-making, indicates, again, the accumulation of songs and music choices that a satellite radio subscription helps to control. Subscription payments provide the subscriber with a sense of ease and purpose among "a consumer culture awash with increasingly ephemeral commodities, saturated by media and advertising," in which "nostalgia has become an increasingly noticeable structure of feeling" to help one find stability.[145] While some find a turn to nostalgia to be the end of progress and innovation, others are careful to complicate nostalgia and pay attention to the multiple meanings it may elicit from listeners.[146] Writing about the simulacrum, art historian Michael Camille argues that Baudrillard's "pessimistic visions arrived at the movement of highest anxiety and nostalgia, simulated ... by new technologies that had totally transformed traditional ways of communication."[147] Media anthropologist Jo Tacchi contends that against claims that nostalgia holds negative social values, it can also be seen to consist of positive social practices.[148] Radio helps to facilitate these practices, Tacchi argues, explaining that the medium's sound is "integrated into daily life in an intimate way and can be understood as forming an important part of domestic environments, or soundscapes, that hold meaning and significance that reaches beyond the immediate context and physical confines of the home."[149] We may also consider how *revivals* of music shape cultural identity and enable music fans to find community and ways of being in popular culture while aging or in enabling a feeling of "generational belonging."[150] One might feel as though the gesture to privilege vinyl on a Classic Vinyl station does service to the era in which they came of age; it works against narratives of technological progress that let older formats or styles of music die off.[151]

A sense of exclusivity through music programming works to grip listeners and justify subscription payments. Certain artists are exceptionally central to SiriusXM's ability to accomplish this, as are practices of expansion and the inclusion of niche genres and audio ambience channels and their nostalgia-tinged sounds. But at the same time that music has been painted as invaluable to the satellite radio companies, it has also served to strategically generate policy and economic wins for the companies, namely through the devaluing of music, a topic I turn to in the next chapter.

4 Disposable Music and Monopoly Power in Policymaking

"Everything dies, baby, that's a fact / But maybe everything that dies someday comes back." In the previous chapter I used this line from Bruce Springsteen's "Atlantic City" (1982), a song that The Band covered and released on the album *Jericho* (1993). The Band's version holds a particular air of melancholy, with the group haunted by substance abuse and the death of co-founding member Richard Manuel in 1986. This 1990s iteration of The Band was minus vocalist, guitarist, and songwriter Robbie Robertson, who received far more royalties and money from the group's songs than the other members. In the American-based music industries, primary songwriters of musical works benefit from the structures that are in place and the policies they shape, much more so than the performers who are not credited as songwriters. Those with songwriting credits profit most handsomely. Copyright and royalty rates are a big part of a framework that largely determines how and when an artist is paid and makes a living. Conversely, they determine how and when an artist will fail to do so.

In the music and radio industries there are ongoing battles to determine royalty rates (the payment one receives for the use of a copyrighted work). Low rates are typically argued for by the media companies that pay them out, like SiriusXM, while artists, rightsholders, and music advocacy groups aim to set rates higher. Over the years, new technologies and methods of music production and distribution have instigated revisions to copyright and royalty frameworks. These revisions have been optimistically anticipated in some cases, given the expanded audiences for musical works made possible by new technologies, but also have been met with skepticism because of concerns about unauthorized copying and piracy, especially in the wake of the internet and file sharing in the late 1990s.[1] Satellites, interestingly, have led some to think that copyright might become irrelevant altogether. A 1964 *Billboard* article on German copyright law outlined a debate around whether melody copyright should be restored to a copyright reform law that had come up before the Bundestag. At the time, the existing law did not allow for the use of any melody taken from a recognized work, but the Bundestag struck the law from the new draft, claiming that it was now outdated. The discussion about the new law referenced satellite technology, stating, "There is some disposition to wonder here whether it is even worthwhile enacting a new copyright law on a national basis, with music now about to become the handmaiden of communication satellite simultaneous worldwide broadcasting."[2] More recently, key acts like the US Digital Millennium Copyright Act of 1998 (DMCA) and the Music Modernization Act of 2018 (MMA) have sought to account for digital technologies and their role in the music business with respect to copyrights and royalties. The reforms they pursue struggle to balance the interests of Big Tech companies, publishers, record labels, superstar artists, smaller and independent artists, and everyday music listeners. An impossible task, to be sure.

This chapter flips the focus of the previous one, which centered on perceived value and music programming, to explore the disposability of music. More specifically, it's the strategic *devaluing* of music that helps a corporation like SiriusXM manage debt, invest profit, acquire talent, and exert monopoly power. The years during which Sirius and XM debuted, as well as those in and around the merger of the two companies, saw change and disruption to record industry models that centered on the production, distribution, and retail of material copies of music, namely CDs. The record business was in "a historic decline." However, digital sales were up 65 percent from 2005 to 2006, and new revenue sources were emerging.[3]

Listeners were consuming as much music as ever, if not more, but analysts claimed record labels were too slow to anticipate the changes brought about, in part, by the internet and digital technologies.

Within the turbulent shifts that technologies and media companies faced throughout the 2000s were artists, the vast majority of whom would see revenue models changing under their feet and who were propelled into the streaming media era, in which the money realized through recorded music would significantly decline.[4] This led some superstar artists to remove their catalogs from streaming services, publicly deride their business models, or refer to music as "disposable" within the distribution and consumption models and practices that these services support. In the early 2010s it was widely reported that listening to streaming music was on the rise. Much less reported was that SiriusXM was increasing its subscriber base after the merger. SiriusXM ended 2010 with just over twenty million subscribers, hitting a company record and becoming the second largest media subscription business at the time, just behind cable company Comcast. The company predicted full-year revenue in 2011 of $3 billion.[5] Despite this growth in music listening for satellite and digital audio services, songwriters were said to be "grossly underpaid" for their music as new technologies shaped the music landscape.[6] In the margins of a February 2014 issue of *Billboard*, two short blurbs about music business news are juxtaposed. On one side, a news story about SiriusXM's increasing revenues uses metaphors of outer space to say that "revenue rockets" and that "revenue went celestial," as it jumped 12 percent from the third to the fourth quarter, from $892 million to $1 billion.[7] Beside this story on SiriusXM is one about Sony Music allegedly withholding royalties from digital downloads to the authors of the 1980s hit song "Eye of the Tiger" by rock band Survivor.[8]

A backdrop of music and media business concentration meant that throughout the 1990s and first decades of the new millennium, major companies were caught up in mergers and acquisitions in efforts to retain or recoup lost profits in the shift to digital and online music products. New deals for royalty rates for digital music were struck that predominantly favored big companies like Sirius and XM as well as the record labels that negotiated their own rates with new market players like Spotify and Apple Music. In 2003 artists, music business executives, and lawmakers testified before the Senate Commerce Committee, arguing that corporate consolidation reduces diversity on radio. Don Henley of the Eagles, Hilary Rosen (president of the RIAA), Senator Russ Feingold, and Representative Howard Berman had gone before the Senate to state that Clear Channel's "ownership of 135 venues and 1,200 stations should be investigated." There were concerns that Clear Channel was using its vast radio ownership to force artists to play at Clear Channel venues.[9] With corporate power in the media industries comes the ability to influence the policy-making processes that shape the cultural industries. As political economy scholar Graham Murdock argues, "All sites of cultural production are fields of contest in which participants with differing stakes and resources jockey for position and advantage."[10]

This chapter argues that a larger context of the political economy of music has shaped the development of the satellite radio industry in North America to mirror trajectories across the music industries that have prioritized big business and shareholders over the livelihoods of working musicians. It focuses on the merger of Sirius and XM, when a key question was whether satellite radio should be part of the same market as terrestrial radio, webcasting, and other digital services, and which technologies it would compete with.[11] In other words, what *was* radio in the digital audio landscape? This brings us back to the metaphor of *streams*. The previous chapter discussed streams to make sense of radio programming in the satellite radio universe, namely with respect to the fluidity of music and its presence in the channel lineup. But another take on this metaphor is the term *streamline*. Following the merger, efforts to streamline resources would work against the more idealistic goals of crafting value through music programming via experimentation and innovation. Next, this chapter discusses how SiriusXM has fought to keep expenses, like royalty payments to musicians and labels, as low as possible. Broader calls to action from advocacy groups and musicians argue that there is room for royalty payments from radio and streaming services to grow, and some progress has been made in establishing new structures for royalty revenue from new media and new companies like SiriusXM. Organizations like SoundExchange and legislation such as the MMA have resulted in some benefits to artists in terms of how much, and by what means, royalties are paid. Through a discussion of the merger and royalty rate negotiations, the chapter makes clear that major media companies like SiriusXM exert significant influence over how music is strategically valued or devalued, which in part determines how and when artists are paid in the wider music ecosystem.

WHEN TWO BECOME ONE: STREAMLINING AND THE MERGER OF SIRIUS AND XM

The term *streamline* evokes rationalization and simplification. It comes up in mergers and acquisitions, where it is argued, often by the merging companies themselves, that to come together is to streamline and to find efficiencies and redundancies to strengthen the bottom line. When the American Sirius and XM merger was first announced, the Canadian national newspaper the *National Post* commented, "One would think that if there were room anywhere for two equally large and vigorous media competitors like Sirius and XM, it would be in the dizzying vastness of outer space." If the companies merged, they would no longer compete for partnerships with automakers, they could streamline channels, and the newfound monopoly power would allow consumers and industry analysts to finally figure out "whether satellite radio is fundamentally a good business idea or a bad one."[12] Some felt that the merger would lead to a monopoly in the already limited satellite radio service industry. To overcome these concerns, Sirius and XM had to prove that they would not only compete with each other, but also with the growing number of digital and online music listening services that were available, or soon to be available, in the mid- to-late 2000s, as well as with broadcast radio.

Before the merger, Sirius and XM experienced financial losses in the quest to acquire subscribers, through marketing and talent acquisition costs. In June 2002 Sirius reported a first quarter net loss of $78.9 million.[13] A few months later, XM's bond prices dropped to 36 cents from 60 cents, a sign that the company could be going bankrupt. Reasons for shaky finances included the costly process of acquiring exclusive deals with automakers and on-air talent, satellite degradations, and the ongoing competition between the two services.[14] In November of the same year, XM grew its subscriber base (adding 64,836 in the third quarter) but was struggling to raise funds. The company was negotiating with automaker GM to convert up to $200 million in payments it owed into debt and convertible securities, contingent upon capital structure changes and the securing of $200 million in financing. XM also cut eighty jobs to make its remaining cash last to the end of the first quarter of 2003.[15] In April 2003 XM and Sirius both increased their revenue, but net losses for the year grew due to higher marketing costs (a loss of $515.9 million for XM and $468.5 million for Sirius in 2002).[16] By the end of June, however, XM and Sirius saw their stock share price increase following successful financial restructurings and subscriber growth.[17] In 2005 the *Wall Street Journal* called Sirius's one million subscribers and $500 million in annual losses a "great start," emphasizing that such companies "grow first and ask questions later." However, the same article also asked, "Do we really need two noncompatible radios so early into this market?"[18] As news of the merger broke, *Billboard* called the companies "bitter rivals" who had grown "tired of the blood sport of their competition."[19]

To complete the merger the companies had to face a congressional hearing as well as weather opposition from organizations like the NAB, whose executive vice president of media relations, Dennis Wharton, argued that the government should not reward two companies that had made bad business decisions with a monopoly.[20] Following the merger's announcement in February 2007, House Judiciary Committee chairman John Conyers formed an antitrust task force. Members of the task force considered how to define the broader marketplace for the digital distribution of music, within which satellite radio operates

In an antitrust hearing in 2007, representatives from the satellite radio industry insisted that the companies competed with "all forms of digital music and retail music and radio."[21] Mel Karmazin, the CEO of Sirius spoke on behalf of both companies, explicitly stating that satellite radio was in competition with both internet radio and broadcast radio. He said, "I can tell you for sure that satellite radio competes with the 10,000 terrestrial radio stations. We compete with over the 1,000 HD radio stations on the air today. We compete with the Internet for Internet radio."[22] To counter Karmazin, David Rehr, the president and CEO of the NAB, was of the opinion that satellite radio was in its own distinct "national, multichannel mobile market," and that the merger would result in a "state-sanctioned monopoly."[23] He framed the service as being in direct opposition to locally oriented terrestrial broadcasting. In response to Rehr, committee ranking member Lamar Smith pointed to the increase in choices and options for receiving music. He believed that satellite radio was in competition not only with broadcast radio but also other programming options that "come without commercial interruption and without the content restrictions that exist on terrestrial radio."[24] Gigi Sohn, president and founder of Public Knowledge, positioned satellite radio as competing with both broadcast radio and digital and mobile music players. She noted that internet radio provided diverse and specialized programming and was becoming more mobile. As a result, internet radio was a viable competitor to XM and Sirius.[25] Numerous claims throughout the policymaking process anticipated increased competition from mobile and online listening devices, suggesting that there was a willingness on behalf of policymakers to approve the merger.

In August 2007 Sirius and XM developed a plan to offer à la carte programming as part of a move to roll out tiered pricing options.[26] The plan was filed with the FCC in hopes that it would help to get the merger approved. The new pricing plan lowered subscription entry prices to $6.99 (with fifty channels) down from $12.95 per month, with the most extensive plans costing $16.99. Karmazin said that these flexible options would only be available if the merger enabled program synergies, or streamlining.[27] The companies had spent millions (Sirius, $5 million, and XM, $8 million) to convince regulators that the merger would "bring unprecedented benefits to consumers and significantly enhance, rather than harm, competition."[28] Built into the logic of the merger, then, were program reductions and cost-cutting strategies that would shape the satellite radio industry as it moved to compete more closely with digital music companies.

The Department of Justice approved the merger in April 2008 after determining that the combined company was in competition with a variety of audio services. In *Empires of Entertainment*, media industries scholar Jennifer Holt uses the term *structural convergence* to describe "the mergers taking place in the media industries from the mid-1980s onward—a mixture of vertical and horizontal integration and conglomeration."[29] Big media companies have strategically widened the definition of their markets to argue that mergers were not instances of consolidation. The SiriusXM merger required final FCC approval from a five-member commission, which had to decide if the deal was in the public's interest. In July the vote on the regulatory body was tied at 2-2, with FCC Democratic commissioners Jonathan Adelstein and Michael Copps voting against the plan. Deborah Taylor Tate's was the final vote, ultimately approving the merger, joining fellow Republicans Kevin Martin and Robert McDowell.[30] The merger involved a three-year price freeze and a commitment to "two dozen channels dedicated to noncommercial programming."[31] On July 25, 2008, the merger was approved by the FCC. Sirius purchased XM for $3.3 billion, not including debt.[32] The commission "determined there was insufficient evidence in the record to predict the likelihood of anticompetitive harms."[33] The FCC claimed that satellite radio was in competition with both broadcast radio and the internet; however, the merger was also indicative of a market logic governing radio, in which any move to improve profits and the viability of a business is considered a sound move. Sirius and XM spent combined merger costs of nearly $150 million, with tens of millions spent on legal fees.[34]

XM's remaining employees were transferred to Sirius, and Sirius provided all the necessary services to XM for it to continue its business.[35] In December 2006 XM had 860 employees;[36] by the end of 2007, Sirius had 973 full-time employees.[37] By December 31, 2008, the merged company had a combined 1,640 full-time employees,[38] which meant a loss of 193 employees, over 10 percent of its workforce, in "synergies." Mel Karmazin became the CEO of SiriusXM, and Gary M. Parsons, former chairman of XM, became the chairman.[39] The new board of twelve directors included Karmazin and Parsons as well as four independent members each designated by Sirius and XM and one representative each from General Motors and American Honda.[40] XM executives who were not given jobs in the new merged company were "assured golden parachute severance packages," some listed at around $5 million and some up to $10 million.[41] Costs in programming, personnel, and essential infrastructure were expected to go down as program redundancies were identified and channel lineups streamlined. Sirius explained that the merger would bring about "net synergies to $425 million for 2009" and that this was expected to increase in years to come.[42] Further, the company's satellite expenditures were expected to "fall off sharply" after 2011.[43] The merged company could now share costly resources and infrastructures, like satellites and terrestrial repeater sites. Heading into the 2010s, it was well-positioned to become one of a few large media companies with massive influence and power in the music industries.

With two services becoming one, channels were canceled due to program overlaps. Those that played music from niche genres were some of the first to go or to move online. In the immediate year following the merger, programming and content expenses decreased slightly, by 1 percent, from $312 million in 2008 to just over $308 million in 2009. This accounted for a full year of XM's expenses being taken on by Sirius but also for savings from "content agreements, personnel and on-air talent costs."[44] XM co-founder Lee Abrams exited the satellite radio industry around the time of the merger but expressed the sentiment that the merger left Sirius with more creative control, and that, in some cases, this meant emulating the terrestrial model. One example of this was reducing the playlist, the overall variety of songs heard on music channels, across the board. And with the merger, Abrams added, the combined companies did not have to compete for acquiring on-air talent; the new company "had a lot of muscle, a lot of power."[45] Labels and artists spoke about what the merger might mean for their airplay and promotion. The majority of record label promotion reps polled by *Billboard* were worried that fewer channels would mean fewer promotion opportunities. Brad Paul, a senior vice president at bluegrass label Rounder Records, was dismayed at losing a national outlet for the label's artists. Those who were involved in promoting artists in niche genres were most disheartened by the news, given the lack of commercial broadcast radio exposure for their artists.[46] In January 2012 *Sing Out!* reported that the folk and traditional music channel The Village would become an online channel. To hear the channel, subscribers needed a premium membership, and the channel was now unavailable in most vehicles.[47]

Though SiriusXM praised the financial soundness of the merger in the trade press and annual reports, the merger did not instantly solve the satellite radio companies' financial woes. Three months after the merger, the new company was "struggling to stay afloat," with three months left to begin paying down over $1 billion in debt, a heavy reliance on automobile sales at a time when the US auto industry was in crisis (requiring a massive government bailout during the Great Recession), and a tumbling stock price that had resulted in an investor lawsuit.[48] In early 2010 questions remained about the viability of the expensive contracts that SiriusXM was responsible for. With stock prices low, *The Hollywood Reporter* asked, "Who's left that can afford to pony up anything near the $100 million a year that Sirius XM has been shelling out to [Howard] Stern and his crew?" Now that the two companies were no longer competing for subscribers, what would this mean for Stern's contract renewal? And did Stern still retain the same cultural cachet he had in the past?[49]

SiriusXM was in talks with EchoStar, DirecTV, and Liberty Media to help circumvent bankruptcy.[50] American mass media company Liberty Media struck a deal with SiriusXM that involved Liberty providing SiriusXM with cash in exchange for a minority stake in the company. This was an investment of up to $530 million in two stages,[51] with a stake as high as 40 percent and an initial agreement that this would not go above 49.9 percent for three years. Liberty Media controlled DirecTV as well, which was the largest US satellite television company. Liberty received board seats proportional to its equity ownership, including one for chairman John Malone. SiriusXM owed $750 million in debt in 2009, and it was anticipated that much of that amount would be "pushed out through waivers, ensuring that the radio home of Howard Stern, Oprah Winfrey and commercial-free music is out of the woods for at least this year."[52] In 2010 Karmazin received a cash bonus of $7 million for "successfully negotiating and executing a restructuring of [the company's] capital structure, including the transactions with Liberty Media."[53] By 2013 SiriusXM boasted twenty-six million subscribers, employed twenty-two hundred staff members, and saw profits of $377 million and $3.8 billion in revenue in that year.[54] The 2013 letter to stockholders proudly stated that SiriusXM was the largest radio company in the world ranked by revenue.[55] Liberty Media had relieved the merged company of financial problems, and over the years its control over SiriusXM would grow. With Liberty's increasing control, SiriusXM not only became a major player in the small pool of large music companies in the streaming media era, but has been able to throw its weight around in policymaking processes that, in part, determine how musicians are paid.

MUSIC AS RATIONALIZATION: "FREE" MUSIC AND STREAMING MEDIA

At the end of a satellite's useful life, the orbiting device is said to be "de-orbited."[56] The idea that a satellite has a useful life appropriately maps onto a popular song and its copyright, with a particular "useful life" when it comes to generating revenue for the rightsholders. Just as some satellites are useful and in operation, others are relegated to disposal orbits, becoming, in effect, space junk. Similarly, some songs are culturally alive and relevant across popular culture, while others are forgotten in basements, garages, and record store dollar bins. Popular music scholars Sarah Baker and Alison Huber allude to "rubbish, trash, waste, junk" when discussing the interviews they conducted with amateur popular music collectors to describe "the anxiety of forgetting or losing culture when popular music artefacts are discarded as 'rubbish.'"[57] What keeps a song or an artist alive in culture versus one that is forgotten? In many ways, media institutions are responsible, as they make choices about what music to circulate.

Describing music as disposable is not unique to the digital age, though it has amplified this notion. The radio industry has used the term *burn* to "refer to a track that has declined in popularity over a period of tests, suggesting that it should now be removed from the playlist."[58] Music industry scholar Mike Jones argues that we know a lot about our emotional attachment to popular music but very little about how music is made. On the production side, Jones locates a logic of failure, pointing to the artists and records that do not meet the criteria required to transform from "original cultural material into mass selling musical commodities." What's left are the accumulated "failed commodities."[59] Jason Toynbee adds that musicians are "constantly being 'dropped' by record and publishing companies. This is an industry which generates failure."[60] A key part of Jones's argument is that it is necessary to partner with key intermediaries in order to generate success, because pop acts cannot be successful on their own,[61] and this is a helpful assertion for thinking about the role of the satellite radio companies in the music industries. A battle for royalties becomes all the more dire when thinking about the record industry as being connected to this notion of failure; royalties remain one way to recoup money long after the flames of initial success are extinguished.

Throughout the first decade of satellite radio service by Sirius, XM, and SiriusXM, questions about the value of music presented themselves in debates about royalties and copyright, particularly as rates were set and introduced for internet streaming radio and streaming interactive services. In late 2005 it was revealed that both companies were paying about 7 percent of their revenues to the music industry. Sirius's year-to-date revenue as of its third-quarter report was just over $160 million, which meant $11 million was paid to the music industry that year, with around $5 million going to labels and artists.[62] As of 2005, the total amount paid to the music industry for broadcast rights from both companies was $80 million.[63] In striking comparison, Sirius had paid $220 million for five years of the NFL and $107.5 million for five years of NASCAR, and XM had paid $650 million for eleven years of baseball. Howard Stern was paid $100 million per year for five years. According to Wall Street analysts covering the satellite radio industry at this time, it was said to be difficult to "pin down a relative value of music versus talk and celebrity programming for satellite radio" because it was never entirely clear what subscribers were listening to.[64]

Music's value, in a context of intangibility and digital distribution, is also informed by cultural assumptions about the value of music, many of which are shaped by changing notions of authenticity, autonomy, and credibility. Over the longer history of recorded music, new technologies and methods of distribution have often disrupted artist income streams. For instance, the introduction of recording studios hurt live musicians' bottom lines, with the number of live musicians earning an income from music falling between 1920 and 1940.[65] More recently, in 2013 Radiohead's Thom Yorke and record producer/musician Nigel Godrich pulled their solo and joint-project music from Spotify because, they said, the company's business model favored large companies and was bad for new music.[66] Daniel Ek, the CEO of Spotify, publicly stated the benefits of a "freemium" revenue model in which music is given away for free;[67] this idea was criticized by many commenters around the time that subscription access to music was becoming normalized in the media industries.[68] Taylor Swift made headlines when she withdrew her music from Spotify and issued a public statement that said Spotify's free tier service undermined the value of music.[69] And below the level of superstar artists like Yorke and Swift, who can afford to pull their catalogs and rally fans around their cause? The vast majority of artists are without such power. In 2018 the average American musician made $21,300 from their music-related work, and the majority reported that music income is not enough to meet living expenses.[70] The digital world and its commercial interactions, promotional culture, and self-branding practices have shifted notions about artists "selling out" and brought music closer to the realm of advertising, corporate sponsorships, social media posts, and television commercials.[71] As music industries scholar Leslie M. Meier argues, the digital age has transformed the core popular music commodity from the sound recording to the artist-brand, with more artists succumbing to music licensing, endorsement, and branding deals and partnerships.[72] These licensing agreements, "arrangements through which music companies sell music to a range of media and brand partners" are increasingly "a crucial complement to the sale of music to end consumers in the contemporary music industries."[73] As with royalties, copyright owners generate income from these agreements.

SiriusXM has strategically devalued music in its business model to aim for lower rates. Record labels have argued that Sirius and XM should pay fairer compensation and stop underpaying music creators, especially given that music is a key reason that listeners subscribe. Of course the record industry has its own bottom line in mind when advocating for its artists, as royalty payments funnel back to labels and publishers, which are often the same company. Labels were also concerned with the low rights fees that XM and Sirius paid at that time, because the new companies were given a financial break while the satellite market was first established. In hearings before the Copyright Royalty Board (CRB) to determine the per-stream rate for the public performance of sound recordings on its internet radio service in 2014, the value of music programming was downplayed by representatives of SiriusXM to help advocate for lower rates. David J. Frear, the chief financial officer of SiriusXM, submitted written testimony that SiriusXM "originally aimed at becoming the world's best music service, but quickly discovered that the ubiquitous availability of music—especially for free on terrestrial radio—required a new business strategy." This new strategy was exclusive and compelling non-music programming that "consumers could not get anywhere else."[74] Non-music programming was said to be what "truly distinguished the service from its competitors and allows the Company to compete vigorously with new market entrants."[75] Stating its approach plainly, Frear added, "Exclusive programming is more valuable to Sirius XM than non-exclusive programming such as music, which is ubiquitous and often available to consumers for free."[76] These comments, made in front of the CRB, gave a lower profile to the value of music in SiriusXM's overall business model for the purpose of keeping royalty rates as low as possible, despite years of promotional materials for Sirius and XM emphasizing the high value of exclusive music programming. Moreover, they reflect a longer history of the radio industry working to circumvent royalty payments to musicians. No doubt such claims by an influential radio company like SiriusXM feed into the larger cultural assumptions about music's decreasing value in the streaming era.

RADIO'S ROYALTY PROBLEM

SiriusXM maintains music programming royalty arrangements with, and pays license fees to, major performance rights organizations (PROs), which collect and distribute performance royalties to songwriters and publishers, including BMI, ASCAP, and SESAC. The company has agreements with these organizations about what the royalty rates are. It also has royalty arrangements with holders of copyrights in sound recordings, which are typically large record companies.[77] SiriusXM also pays royalties to SoundExchange for digital public performance of sound recordings, at rates that are negotiated through the CRB. Unlike American terrestrial broadcast radio, SiriusXM pays public performance rights royalties, which are payments to performers who are not credited as songwriters.[78] Copyright legislation crafted in 1972 in the US accounted for physical sound recordings and included a federal copyright for sound recordings that compensated artists and composers for the reproduction and distribution of musical works. But the Sound Recording Act of 1971 failed to protect rights in the public performance of a song, which includes songs played on the radio.[79] The radio industry lobbied to have the performance right excluded from the 1971 act, based on the argument that the radio provides promotional benefits to artists and labels.[80] Further, American artists do not receive performance right payouts from other countries because artists outside the US are not compensated when their work is played on American radio. The DMCA introduced a performance right for digital audio, which incited heavy debate between the RIAA and broadcasters over whether a simultaneous webcast of an AM or FM broadcast made the broadcaster liable for the copyright.[81]

To this day, music advocacy groups are pushing for the collection of performance royalties for AM and FM radio airplay.[82] One thing that sparked further interest in advocacy to establish a performance right for terrestrial radio was the fact that newer music platforms, like satellite radio, do indeed compensate artists for the use of their music when performed on the primary satellite service as well as the simulcast streaming.[83] If SiriusXM pays, many thought, why shouldn't broadcasters? In August 2007 folk singer Judy Collins appeared before Congress to argue this very point. The MusicFIRST coalition was formed by more than 150 recording artists and other unions and organizations to remove the public performance right exemption for terrestrial radio.[84] Against this, some smaller radio stations have claimed that they would be put out of business if they had to pay more royalties to performers.[85] Throughout these debates and discussions, artists have been vocal about the fact that record labels have a tendency to "keep whatever money they do collect for themselves—at the expense of musicians who have no choice but to sign increasingly stringent contracts."[86] One way that performance royalties have been established in terrestrial radio is through radio companies making direct licensing deals with labels. In 2012 Clear Channel struck a deal with country music label Big Machine to pay a recording performance royalty, to be split equally between labels and artists. But the NAB continued to fight similar moves across the radio industry on the basis that radio still promotes music and doesn't replace sales.[87] The next year, Warner Music Group became the first major label to receive performance royalties for American broadcast radio airplay through an agreement with Clear Channel.[88] Yet the work is ongoing to have these rates be solidified in policy.

One example of a rights organization that has helped to funnel royalty payments to artists is SoundExchange. SoundExchange is a nonprofit subsidiary of the RIAA and was appointed by the Copyright Office to license and collect royalties from noninteractive digital services, which include cable radio, satellite radio, and webcasters. The payments that

SoundExchange collects are sent directly to featured artists and sound recording owners who have signed up with the organization.[89] It was founded in September 2003 after efforts to introduce a public performance right and a mechanism for artist compensation in the wake of downloading and file sharing. In 2004 the group represented over eight hundred record companies and thousands of artists.[90] Both The Digital Performance in Sound Recording Act of 1995 and the DMCA gave owners of musical works "the exclusive license to perform the copyrighted song publicly by means of digital audio transmission," which enabled the emergence and growth of music streaming services.[91] In 1995, the Digital Performance Right in Sound Recordings Act (DPRA) indicated that subscription-based, non-interactive internet transmissions, those that operate similarly to a radio broadcast, would pay a statutory license fee set by the CRB, and that interactive internet transmissions services like Spotify would negotiate licenses with copyright holders.[92]

In 2004 XM and Sirius were included in SoundExchange royalty payments for the first time.[93] Ann Chaitovitz, an artist advocate with the American Federation of Television and Radio Artists (AFTRA), gave the example of the song "Papa Was a Rollin' Stone" by the Temptations to distinguish between the payouts via satellite radio and broadcast radio. If it was heard on broadcast radio, the songwriters (Barrett Strong and Norman Whitfield) and the publisher would be paid, but the Temptations and Motown would not. If it was heard on XM, the songwriters, the publisher, the Temptations, and Motown would all be paid.[94] SoundExchange's fall 2004 allocation involved $6.5 million in sound recording performance royalties to artists and record companies, and the collection group anticipated that much more would be on the way over the next few years due to the growth of satellite radio and webcasting.[95]

Across much of the music industries, a sense of hopefulness was expressed with respect to the SoundExchange royalty system. Because of meager payouts from on-demand streaming services, and given that broadcast radio in the US only pays royalties to songwriters and publishers, not performers or labels, the revenue stream was thought to be a boon to artists' careers. The RIAA had started SoundExchange to collect and distribute these fees, and artist advocacy groups worked together over three years to guarantee SoundExchange would be an independent organization controlled by both artists and labels, and that artist money would go directly to artists, not through labels.[96] In October 2013 performance royalty distributions had grown significantly thanks to SoundExchange, with anticipated royalty distributions of $500 million that year. SoundExchange took a 4.9 percent administrative fee, and then 50 percent of the remaining amount would go to owners of the sound recordings, 45 percent to performing artists, and 5 percent to non-featured artists through the American Federation of Musicians and the Screen Actors Guild-American Federation of Television and Radio Artists.[97] By the end of 2018, SoundExchange's distribution payouts had grown to $953 million;[98] SiriusXM also boasted record revenue of $5.77 billion in that year.[99]

With SoundExchange and the performance royalty in effect for satellite radio services, a new revenue stream was established for labels and artists, which allocates the lion's share directly to artists. However, the satellite radio companies have consistently sought to keep these rates as low as possible, often citing the fact that terrestrial broadcast radio does not pay them as a reason they should not increase quickly or substantially. In 2006 negotiations were in progress for rates for the period between 2006 and 2010. Sirius and XM had made a deal with the recording industry in 2002, reportedly for $80 million, which was also expiring in 2006. Labels were now hoping to get the sort of money that Sirius and XM had paid for sports and entertainment companies and big name brands.[100] In December 2006 the CRB set the royalty rate for XM and Sirius for a six-year period beginning January 2007 and ending December 31, 2012, which had the newly merged company paying a performance license rate of 6 percent of gross revenues subject to fees for 2007 and 2008, 6.5 percent for 2009, 7 percent for 2010, 7.5 percent for 2011, and 8 percent for 2012.[101] Because around half of SiriusXM's programming is talk radio, the percentage applied to about half of the company's revenue. The effect of these rate increases was routinely bemoaned in company annual reports. In 2007 Sirius spent close to $147 million on revenue share and royalties, an increase of 110 percent from 2006's amount of nearly $70 million. The company explained that this increase was due to the new royalty rate of 6 percent of gross revenues, a new expense of $48 million, as well as the increase in subscribers over this year.[102] These expenses were said to be tough to manage if they were to get too high, likely part of a larger rhetorical strategy to sway rate negotiations with the CRB in the company's favor in years to come.

SiriusXM has also worked to bypass SoundExchange by establishing direct licenses with labels. The company said this would enable subscribers to have more control over programming, such as the ability to record programming,[103] a feature it had been working to establish and sustain, and this would allow SiriusXM to compete more directly with on-demand streaming platforms. In 2011 SiriusXM was proposing a rate of 7 percent under the direct licenses it pursued, 1 percent lower than what its CRB rate would be in 2012. Some artists expressed concern that royalties would be delayed or smaller if they were to come through labels as opposed to through SoundExchange. Services like Spotify and Pandora, with more interactive features, however, paid higher rates to the industry.[104] In 2012 the formula for Pandora paid either 25 percent of revenue or 0.102 cents per play, whichever was greater; the latter in 2011 was 50 percent of revenue.[105] With services like Spotify and Pandora, the money moved through direct licenses with record labels and their contracts with artists, so the payout per play was much lower than for a SiriusXM play. Satellite radio also paid more to artists because each play reached every subscriber hearing the song at that time, as opposed to the one person who would be listening to an interactive streaming service. There are only so many times a song can be played on satellite radio. Each play counts for more.[106]

The CRB set SiriusXM's performance royalty rate for 2013 at 9 percent of gross revenue, which some analysts argued was too low, "by several percentage points, depriving artists and labels of tens of millions of dollars of royalty payments which will instead flow into SiriusXM's coffers."[107] In other words, the growth of satellite radio was argued to have been "artist-subsidized." Before this rate was set, SiriusXM filed a request with the CRB in November 2011 for a royalty rate of less than 7 percent of gross revenues, subject to exclusions. By comparison, SoundExchange filed a request that the rate begin at 12 percent, increasing by 2 percent each year up to a maximum of 20 percent of gross revenues.[108] Between 2009 and 2012, the merged company's revenue increased from under $2.5 billion to over $3.4 billion, and former CEO Karmazin said, "If we want a performer . . . we can afford to pay more than anybody else can because we're making more."[109] SiriusXM's rate increased to 11 percent in 2017, an increase phased in over the license period. The following five-year period, beginning on January 1, 2018, saw the rate set at 15.5 percent of gross revenues.[110] This was a significant increase from the 11 percent it was previously paying but much less than the 23 percent that SoundExchange was asking for at that time.[111] Although SoundExchange was dismayed at the 15.5 percent rate, *Billboard* wrote that SiriusXM still had "one of the highest per play rates in the U.S. industry, paying some $25–$30 per play."[112] These rates, however, could be reduced each month due to songs being played with direct licenses from copyright owners and due to the percentage of songs played that were recorded before February 15, 1972.

In the 2013 to 2017 license rate period, SiriusXM was hit with a number of lawsuits regarding its payment, or lack of payment, to artists and labels. One widely publicized lawsuit was filed by the Los Angeles folk rock band the Turtles, known for hits like "Happy Together" (1967), which sought artist performance royalties from SiriusXM. The lawsuit was part of a larger issue that involved SiriusXM skirting payments for recordings with pre-1972 copyrights. As mentioned earlier, a nationwide US copyright was created for master recordings beginning on February 15, 1972, and thus SiriusXM argued that it did not have to pay for recordings made before this date.[113] For songs recorded before the 1972 copyright, SiriusXM was deducting about 10 to 15 percent of gross revenue, from the 50 percent or so that it paid royalties on.[114] On September 22, 2014, a California judge ruled that SiriusXM had violated the public performance rights of the Turtles and was liable for infringing ownership rights.[115] This ruling was only for California, but it was said to potentially set legal precedent for other states to make similar decisions with respect to their own copyright laws. Even though the ruling was just for California, listeners in the state could not be excluded from the national satellite service, so the company would need to pay the royalties, stop playing the music, or create a different state-specific payment formula.

SiriusXM appealed the decision and went as far as to threaten to remove the pre-1972 works from its service.[116] But in 2015 the company struck a $210 million settlement with record companies for its use of pre-1972 sound recordings. In November 2016 SiriusXM also settled with the Turtles, for an amount between $25 and $40 million. The band won its suits in California and New York but lost in Florida. The settlement meant that the Turtles and independent labels that owned music created before 1972 would receive royalties from a pool of at least $25 million, and SiriusXM was given a ten-year license to play the pre-1972 recordings.[117] In February 2017 Sirius won dismissal of the New York copyright lawsuit because New York copyright law did "not protect the public performance of songs made before 1972," which reduced the size of the settlement from November.[118] Major labels were unhappy with the ten-year license royalty rate of 5.5 percent for pre-1972 songs, claiming that a "10-year license of any sort 'is virtually unheard of in the music industry, where new platforms, technologies, and business models are constantly emerging,'" and that the 5.5 percent rate was a far cry from the then-current rate of 11 percent being paid by SiriusXM for all other songs.[119] SiriusXM defended the rate by stating that no recording owner had ever asked any broadcaster to stop performing pre-1972 recordings before this lawsuit, and that the negotiated royalty rate exceeded the highest rate paid to any of the independent labels that had entered into direct license deals with SiriusXM covering pre-1972 recordings.[120]

SiriusXM continued to fight rate increases in hopes of keeping its costs low and to return dividends to shareholders and continually invest in celebrity radio talent to maintain exclusivity. The merged satellite radio company was termed "Music's New Best Frenemy" by *Billboard*, given its complex position in the music industries. On the one hand, it had "paid record labels and artists hundreds of millions of dollars annually while promoting unsigned and emerging acts on its niche music channels." On the other hand, its relationship with the music business had "grown more contentious" with its opposition to paying pre-1972 royalties and its efforts to establish direct licensing deals.[121] Eventually, SiriusXM would become aligned more closely with its streaming music competitors in the copyright and royalty framework, thanks to new legislation that aimed to compensate artists more fairly for their recordings.

THE MUSIC MODERNIZATION ACT

In 2018 the usually fractured music business saw various components come together to mobilize against the power and influence of Silicon

Valley.[122] Several bills were introduced in both the House and the Senate that year, the result of a music industry movement that had been growing over the previous decade, but songwriters, artists, publishers, producers, and distributors had organized behind three bills: (1) the Music Modernization Act (MMA), "which among other things, would create a mechanical licensing collective for all digital music, so that streaming entities can find and compensate artists for their work"; (2) the Compensating Legacy Artists for Their Songs, Service and Important Contributions to Society Act (CLASSICS Act), "which would add copyright protection for pre-1972 sound recordings"; and (3) the Allocation for Music Producers Act (AMP Act), "which would allow for the payment of performance royalties to producers, mixers and sound engineers of sound recordings."[123] In advance of the Grammys on the Hill event of 2018, the House Judiciary Committee chairman, Bob Goodlatte, put forth a new piece of legislation that combined elements of these three bills. This new bill would "create a new organization to collect and distribute mechanical royalties, change how much some digital services pay to use recordings, require them to pay for those made before 1972 and codify the process by which SoundExchange pays producers and engineers directly."[124] Dubbed the MMA, the consolidated bill passed in the House of Representatives with a vote of 415–0 on April 25, 2018. It still needed to pass a Senate vote, and SiriusXM was adding more lobbyists to its fight against the need to pay royalties for pre-1972 recordings.[125]

During the committees and hearings that preceded Senate approval of the MMA, companies like SiriusXM and artists, such as Smokey Robinson, offered their perspectives on what this legislation would mean for industry operations as well as the livelihoods of recording artists. Submissions were made to the US Senate Committee on the Judiciary Hearing on Protecting and Promoting Music Creation for the 21st Century on May 15, 2018. One of the more significant topics of discussion was a last minute change to the 801(b) standard, which sought its removal and replacement with the "willing buyer, willing seller" standard for all companies streaming music online. The MMA defines this willing buyer, willing seller standard as a process by which the CRB establishes "rates and terms that most clearly represent the rates and terms that would have been negotiated in the marketplace between a willing buyer and a willing seller."

This includes taking into consideration whether the service substitutes or promotes the sale of records or if it might interfere with or enhance other revenue streams of the copyright owner, the relative roles of the copyright owner and the transmitting entity in the copyrighted work and service made available to public with respect to creative contribution and cost, and the rates and terms for comparable types of audio transmission services and comparable circumstances under voluntary license agreements.[126] The 801(b) standard, which at the time set the rates for SiriusXM, determined rates by attempting to balance public and corporate interests, aiming to minimize disruption to the company.

Along with SiriusXM, the "world's first and oldest digital music service," Music Choice, claimed that last minute provisions were "grafted" onto the MMA, namely the implementation of the willing buyer, willing seller standard, and there was no input from the three non-interactive services it would affect, including SiriusXM.[127] Music Choice argued that a one-size-fits-all solution was inappropriate given differences between all the services affected by the MMA and because terrestrial radio still did not pay performance royalties. Also, because most of the major parties who participated in the MMA negotiation process maintained direct licenses with recording companies, they would not be subject to the rules of the willing buyer, willing seller standard. Music Choice said that the new standard had not proven successful for any company, with most going out of business or never becoming profitable. A list of defunct companies, or a "Willing Buyer/Willing Seller Graveyard," provided by Music Choice, included AOL Radio, Beats Music, Last.fm, NetRadio, Radio.com, Yahoo! Music, Songza, and Pandora. Pandora was still in operation but had transitioned to on-demand and direct licensing.[128]

Against the points raised by SiriusXM and Music Choice, Mitch Glazier, president of the RIAA, proclaimed that the MMA was "a consensus bill that has been debated and refined over the course of several years and has garnered widespread support from both the music and technology industries."[129] Glazier spoke positively about its components, praising its ability to fairly compensate legacy artists, establish an effective royalty and payment system for music producers, and ensure all creators were compensated at market-based rates across all digital platforms. Chris Harrison, the CEO of the Digital Media Association (DiMA), testified as well, and his submission indicated how the music and tech industries came together to advocate for the bill. DiMA included members like Amazon, Apple, Google, Pandora, Spotify, and YouTube, but not the non-interactive internet radio services like SiriusXM or Music Choice. These companies were in favor of creating a "true blanket license" of the sort that Music Choice spoke out against.[130]

In addition to representatives of media companies and industry representatives, legendary recording artist Smokey Robinson testified at the hearing, and his words centered on the benefits of the policy changes introduced by the CLASSICS Act. Robinson proclaimed, "Musicians who recorded before February 15, 1972 deserve to be compensated the same way as those who recorded after that date." He added that these pre-1972 songs were indeed *classics*, that they resonate today and "add value to our lives and bring people together. They define America. And their financial value to the companies that play these recordings is clear." Robinson highlighted channels like SiriusXM's 50s and 60s decades channels, which could play classic songs but not have to pay the performance rights to the artists who made the music. This made these channels especially lucrative for the satellite radio company, given the financial breaks they provided. As Robinson said, the "companies that exploit our music for profit should pay us for it—pure and simple."[131] This echoed earlier comments made by Mary Wilson of The Supremes. Wilson said that songs "the group recorded in the 1960s, like 'Stop! In the Name of Love' and 'Baby Love' are treated with 'less value' by SiriusXM channels than their 1976 hit 'I'm Gonna Let My Heart Do the Walking.'"[132]

SiriusXM opposed the CLASSICS Act because, as the company said, it did not cover terrestrial broadcast radio. Four days before the Senate vote scheduled for September 14, SiriusXM stepped up its objection to aspects of the MMA after learning that some compromises had been made to the NAB and television-based music service Music Choice in exchange for these organizations' support for the bill.[133] In quick response, an op-ed by attorney and MMA advocate Dina LaPolt called for the board chair of Live Nation and SiriusXM Holdings, Inc., Greg Maffei, to resign.[134] However, Susan Genco, the co-president of the Global Music Rights PRO, negotiated a compromise with SiriusXM and the Big 3 record labels of Universal Music Group, Sony Music Entertainment, and Warner Music

Group. Negotiations continued up until the moment right before the vote on September 18, and the bill then passed unanimously.[135] SiriusXM fought the MMA until "labels agreed to lock in its current 15.5 percent-of-revenue royalty rate until 2027."[136] Recall that SiriusXM had previously been given a 15.5 percent rate by the CRB for the period of 2018 to 2022. SiriusXM also agreed to an amendment that guaranteed artists would be paid 50 percent of the money that the company paid to record labels for pre-1972 sound recordings.[137] With the passing of the MMA, CEO of SiriusXM James E. Meyer, who replaced Karmazin in 2012, said that the official SiriusXM attitude was that "stability is good," referring to the fact that its rate was locked in for the following nine years.[138] The passing of the MMA also meant that the new willing buyer, willing seller standard would replace the 801(b) standard for SiriusXM following this nine-year period.[139] However, SiriusXM had been eyeing direct deals with labels, which cut out administrative fees and would circumvent SoundExchange. As of 2016, SiriusXM had made deals with around five hundred independent labels that made up around 6 percent of the overall songs it played.[140] By establishing more direct deals with labels, the wins for artists gained by SoundExchange could be diminished. This is a major issue in SiriusXM's negotiations with the CRB and with labels in years to come and one that exemplifies the ongoing strategies deployed by major media institutions trying to bend the rules in their favor.

What these hearings and policymaking procedures indicate is that SiriusXM is one of a few major players to wield massive influence on legislation. The vast majority of artists are not present or active in these processes, and they are then left to play by the rules set by big players. SiriusXM's ongoing efforts to maintain low royalty rates shed light on the company's longevity and success over decades during which countless music streaming services have started and shuttered. The grandfathered 801(b) standard has helped SiriusXM to grow and sustain itself, as has its monumental merger and the influx of cash it received from Liberty Media. The company's institutional power plays into policymaking discourse, in which the company has squared off against broadcasters and artist advocacy groups. At the same time, SiriusXM has been one of the highest providers of royalty payments that artists receive, thanks to its performance right payments, the SoundExchange system in place, and the large number of listeners and subscribers who use the service. In fact, a number of Canadian artists have said that SoundExchange royalties from SiriusXM are some of the most substantial they receive.[141] These issues are complex, and often the competition that takes place between major companies and organizations leaves artists in the shadows, left to make do with the results of debates, discourse, and policies set in boardrooms and by governments swayed by the influence and power of big business.

What this chapter, and the previous one, have indicated is that value and disposability can be crafted and constructed in strategic ways. On the one hand, perceived value can serve to entice and retain subscribers, where a distinct style of music programming becomes the selling point for a listener to become a subscriber. Niche music channels speak to their own tastes and subjectivities in ways that other music services and commercial broadcast radio do not. On the other hand, music can be downplayed by SiriusXM representatives in order to lower rates, particularly in comparison to big name talk show hosts and branded channels that have received massive investments by the satellite radio company. The next chapter picks up on these issues and pays closer attention to specific channels and radio hosts to get a sense of how music reaches listeners in intimate ways, especially through the influence of celebrity musicians as radio hosts who connect with subscribers to further demonstrate the exclusivity and value of a satellite radio subscription.

The satellite radio companies are just one part of the broader issues and debates surrounding artist compensation in the new millennium. In 2014 Beats Music CEO Ian Rogers claimed that competing services have devalued music in their pursuit of high subscriber numbers through lower subscription rates. Rogers said, "We have massive services out there that are training consumers to think that music should be free at a time when we have a hundred million people in the US paying for cable or satellite [TV]."[142] Some artists have challenged YouTube's minuscule royalty payments to artists as well as the hosting of unauthorized videos. In 2016 the music business was said to have "less bargaining power than ever" against the increasing power of platforms like YouTube, which 98 percent of internet users between the ages of eighteen and twenty-four visit regularly. Irving Azoff, manager of bands like the Eagles and Van Halen, called YouTube's revenue for artists "a joke" and said that the company acts "like an old record company by making the accountancy so difficult, the artist remains in the dark."[143] Spotify similarly came under fire for unlawfully distributing copyrighted musical compositions and was sued in 2015 for doing so.[144] Along with companies like Spotify and PROs, SiriusXM has strategically navigated hearings and legal proceedings in a way that ensures the security of key deals with major labels and individuals, essential to the service's profitability and bottom line. These seemingly status quo power dynamics are succinctly and emotively communicated in American country and folk singer-songwriter Gillian Welch's haunting and eloquent "Everything Is Free," in which she sings, "But everything is free now, / That's what they say. / Everything I ever done, / Gonna give it away. / Someone hit the big score, / They figured it out. / That we're gonna do it anyway, even if it doesn't pay." A theme of freedom, one often tied up in music's more revolutionary aspects, is now also suggestive of the heavy lamentation of the decreasing value of musicians' work in the streaming age.

5 The Stars Down to Earth

THE CELEBRITY RADIO VOICE
ON SATELLITE RADIO

The Beatles entered homes across the world with the *Our World* two-hour live television show in 1967. During the program, the band's upcoming single "All You Need Is Love" was played for an estimated audience of five hundred million in thirty-one countries. This major television event was made possible by three American communications satellites and one Russian satellite.[1] Decades later, satellites continued to facilitate these connections between stars and audiences. In January 1995 Pearl Jam, a band now with its own SiriusXM channel, broadcast a four-and-one-half-hour live show on satellite airtime purchased by the band, which was made available to any radio station that wished to carry it. The live show was followed by frontman Eddie Vedder playing "his favorite records for an hour."[2] The satellites, in effect, brought these stars down to earth and into the living rooms of people around the globe.

The satellite, and its values of expansive geographic range and musical variety, has helped to connect stars with listeners across time and space. After Sirius and XM debuted, some analysts felt the time was right to return to "personality music radio" as a means to create excitement about music. One 2005 op-ed in *Billboard* argued that the power and prominence of the DJ had lessened over the years, and it was time for the figure to return. It pointed to the success of talk radio and referenced Sirius's hiring of Howard Stern as an example of how to bring listeners to radio. There was room for this sort of on-air personality for music radio, the op-ed argued.[3] Against the automation, standardization, and syndication of commercial radio programming, a nexus criticized by many radio listeners, the figure of the radio host, and the human element it brings to the airwaves, has enabled music radio to intimately connect with listeners. Importantly, the use of prominent personalities to shape music programming and entice listeners is not exclusive to satellite radio.[4] As of 1926, Duke Ellington was a feature of the airwaves; in 1948 he "found an opportunity in the short-lived celebrity disc-jockey fad." As music and media scholar Aaron J. Johnson writes, "Ellington was on the radio because he drew listeners."[5] This human element is an indispensable component of satellite radio programming. SiriusXM is especially strategic in targeting big name celebrities and those with loyal fan bases to maintain its exclusivity.

This chapter investigates the role of celebrity talent in ascribing value to satellite radio music programming. SiriusXM's ongoing investment in talent has been regularly touted as a key aspect of its success. In early 2013, at a time when SiriusXM was just behind Netflix in total number of American subscribers, 23.9 million to Netflix's 25.1 million, and well beyond Spotify's 1 million, *Billboard* wrote, "No one wants to pay $10 per month to find needles in haystacks."[6] This assertion suggests that SiriusXM's personal and curatorial elements have given the subscription radio company an advantage over some of its competitors. As earlier chapters indicate, talent costs money, and SiriusXM has worked to find a balance between hosted and hostless channels in order to appease those subscribers who prefer a hosted channel but also avoid the use of a host on channels where a live human presence may not be deemed necessary by listeners. A central aspect of SiriusXM's subscriber business model is to present the radio service's value through celebrity personalities and the social, authoritative, and conversational voice that they use to animate and annotate music radio. While there are many hosts on the channel lineup, three central archetypes heard on the music channels are the *collector*, the *tastemaker*, and the *veteran*. Each has its own ways of contextualizing and accrediting music. Often they inspire listeners to find and listen to and purchase music in other venues or aspects of the music industries. Voices thus *pilot* the listening experience and propel the cultural industries through tuning listeners in to music and shaping their purchasing decisions.

RADIO AND SOCIABILITY

Radio's social function, or its *sociability*, has been a long-running and defining feature of the medium. Sociability, according to Paddy Scannell, "is the most fundamental characteristic of broadcasting's communicative ethos."[7] It also has some radio scholars arguing that newer audio formats like podcasting are indeed worth considering *as radio* due to its ongoing prominence. Radio and sound scholar Andrew Bottomley says that "radio as a medium has stayed relatively consistent across its century-long history, if and when it is viewed principally as a set of cultural relations instead of as a technology, an industry, or an aesthetic formation alone. And it is the characteristic of sociability, rooted in the element of liveness, that creates the social space that makes radio *radio*, as an aesthetic formation."[8] Writing about the space between the production and reception of transmitted voices, John Durham Peters explains that "intimate sound spaces, domestic genres, cozy speech styles, and radio personalities all helped to bridge the address gap in radio." Those who succeeded in the medium's early years, specifically the "golden age" of the 1930s, were "crooners, comics, and avuncular politicians, people who knew how to 'reach out and touch' their audiences."[9] Whether in song or in mediated voices on the radio, listeners can hear their lives in these voices. Simon Frith makes this point in reference to singing, in which we hear life "despite and not because of the singer's craft, a voice that says who they really are."[10] I believe this is the case for radio voices as well. As radio scholar Kate Lacey argues, ideas and debates about intimacy and the simulation of presence in the social media era can be traced back to the early years of radio broadcasting, "where liveness, immediacy, voice, perpetual connection and universal access already characterized the form."[11] Radio has a long history of facilitating community formation through the act of listening to voices.

Through conversation and the playing of music, the radio host constructs meaning for the listener and fosters a sense of connection.[12] Radio scholar Chris Priestman argues that there is, on the one hand, radio that fosters conversation and then, on the other hand, radio that is limited to music distribution, that which is "automated, hostless, and pre-programmed."[13] Further, in Priestman's analysis of radio narrowcasting and online conversation, the use of online message boards during the early years of XM Radio was one place where a public forum for satellite radio users developed, and where conversation about music and radio flourished. In these posts, users regularly championed the radio host on XM. The host lends a personal feel to music by comparison to the automation of FM radio and the impersonal feeling of MP3 playlists.[14] Jeffrey Roessner adds that the radio host on satellite radio propels its "anti-commercial aesthetic," free from demands of advertisers and able to offer more spontaneous and free-flowing talk by presenters. But because of the subscription fee required to access programming, "SiriusXM has, paradoxically, commodified the anti-consumer aesthetic" of radio forms like American freeform FM or public and independent radio.[15] Thus, Sirius and XM's emphasis on the value of "personality" radio is another instance of restaging aspects of noncommercial broadcast radio, namely freeform, college, and community radio, which have long communicated music's significance via knowledgeable radio hosts.

The radio voice is also a means by which listeners navigate the seemingly endless choices in the digital era. Within the extensive channel variety and wide geographic coverage of satellite radio, the voice of the radio DJ accompanies music in meaningful ways, and music is given a sense of value when the voice belongs to a celebrity.[16] The host uses their expertise to curate music selections, not merely play them. As Simon Reynolds claims, the act of selecting music for radio or bands for a festival has now been painted with the term *curation*, a "high-falutin' gloss," as he says, when before this would have been described as "selecting" or "booking."[17] There is a dismissive tone to Reynolds's discussion of the seemingly ubiquitous use of the term curation, which to him has nearly lost all meaning, and the term serves to easily apply cultural capital, or again, perceived value, to the simple task of selecting and accumulating; this is made to seem more prestigious when the one doing the curating is a celebrity.

THE STARS ON AIR

Connecting celebrities to radio channels has always been a key part of Sirius's and XM's programming strategies, but the use of celebrity hosts has grown and intensified over the years. Early on, artists and key figures in music developed programming for XM, including record producer Quincy Jones and jazz musician Wynton Marsalis. According to XM's Lee Abrams, the company was after "real, authentic artists," those who "were timeless and would be around in the next 20 years and really have something to say." In some cases, such as with Bob Dylan, it would take a lot of effort to secure the partnership, while in others, such as with Tom Petty, it was a quick "yes."[18] The early Sirius channel lineups were thin on celebrity talent, with no music channels devoted to any one band or artist, but this would change in years to come. For example, Sirius launched Eminem's Shade 45 channel in 2004 following the artist's massive success with the best-selling album of 2002, *The Eminem Show*.

Before channels were created around a single band or artist, celebrity musicians hosted programs that ran periodically or helped produce specific channels. In 2000, Queen Latifah signed a three-year deal with Sirius, as had Grandmaster Flash, Sting, and BeBe Winans.[19] Country Music Hall of Famer Bill Anderson had a contract with XM, and "old-school" hip hop DJs and MCs were hosting on XM, such as producer and Cold Chillin' Records founder Marley Marl.[20] Snoop Dogg was promoting XM in 2003 as the executive producer of XM's classic hip hop channel The Rhyme.[21] By 2004 XM was leaning heavily into exclusive, artist-based programming, introducing shows like *Artist Confidential*, which featured one-on-one interviews and performances from "music legends" like Bonnie Raitt, Don Henley, Brian Wilson, and Emmylou Harris.[22] One *Artist Confidential* show that featured Paul McCartney was "very successful," and McCartney brought up an audience member to play piano and write a song on the spot. It took around six months to land McCartney for the program.[23] In early 2004 Sirius was "upping its profile in the music industry by recruiting several celebrities for various programming and promotions," including LeAnn Rimes performing on the Sirius Stage at the Consumer Electronics Show in Las Vegas and the signing of "Little Steven" Van Zandt to develop the "24-hour garage-band stream" Underground Garage.[24] In 2007 Sirius recited programming accolades that showcased its growing roster of talent. These included launching Siriusly Sinatra (a channel featuring music from Frank Sinatra and related artists), debuting The Grateful Dead Channel, bringing back E Street Radio (to coincide with Bruce Springsteen and the E Street Band's 2007 album *Magic*), and launching limited time channels with Duran Duran and Garth Brooks.[25] Acquiring celebrity talent entices subscribers. An artist-based channel encompasses voice, talk, and the more intimate details from these celebrities' lives. Subscription access entails access to one's favorite music artists.

After the merger of Sirius and XM, the use of celebrity-hosted shows and artist-based channels was a continuing pillar of the satellite radio business model. In 2013, in a profile of Scott Greinstein, president and chief content officer for SiriusXM, he was asked, "Why are so many major artists—from Bruce Springsteen and Bob Dylan to Tim McGraw and Eminem—willing to partner on original content?" Part of the answer, according to Greenstein, was a programming philosophy that allowed one to "program as human beings, not algorithms" and that "artists trust what we do because we don't have an agenda." Greenstein called this "handcrafted" programming, and these programs would build trust with artists and listeners and drive top-line revenue.[26] A year later, Greenstein echoed this sentiment and said that his approach had been to bring celebrities "from all walks of life" to the radio, regardless of their prior broadcasting experience. He cited the examples of Marky Ramone, Tony Hawk, Bruce Springsteen, Eminem, Barbara Walters, Phil Jackson, and Nancy Sinatra, and said that they all turned out to be interesting hosts. Greenstein added that he let "talent and content do their thing and then figure[d] out how to market it. . . . The creative process should be as *pure* as possible'" (emphasis added).[27] The previously mentioned archetypes communicate the service's perceived value, bridge the gap between an artist or band's recorded music output and fan communities, deliver an already established fan base to the subscription service, and appeal to a subscriber's desire to hear music that isn't widely available on broadcast radio, or that is curated in ways distinct from algorithm-driven music platforms.

THE COLLECTOR

One prominent radio figure, particularly in rock and rock-derived genres and formats, is the *collector*-as-expert: a musician who is known for longevity in their career as well as having a perceived deep knowledge of music. These hosts boast immense star power and typically have crafted and controlled a notion of authenticity that shapes their star text. In other words, listeners are already convinced of their expertise. Tom Petty, Bob Dylan, and Bruce Springsteen are examples of musicians who fit this archetype, and each has had a channel or a radio show organized around their music. Due to the ideological and stereotypical connections between authenticity and masculinity in the rock genre, pertaining to race and gender, collectors-as-experts have often been male and white. Outside of rock, celebrity musicians like B. B. King and LL Cool J also embody aspects of this archetype, as they come to represent genres like blues and hip hop on the channel lineup. B. B. King hosted a program on his own channel, *B. B. King's Bluesville*, which began in 2008. He was dubbed the "Mayor of Bluesville," and he added stories and discussion to the songs he played. Later in this chapter, LL Cool J is discussed as communicating aspects of both the veteran and collector archetypes. The collector archetype also exemplifies the earlier years of music programming on Sirius and XM, when legacy acts promoted ideas of prestige radio through deep dives into music programming that would go beyond the playlists of commercial broadcast radio.

The figure of the collector is notably present in Bob Dylan's acclaimed XM radio show *Theme Time Radio Hour* (*TTRH*), which ran on the Deep Tracks channel from 2006 to 2009. Each episode of *TTRH* ran for about an hour (or sometimes just over). There are a total of three seasons and 101 episodes (100 episodes were aired, but a "lost" episode was later discovered in 2015 and was broadcast on SiriusXM, the BBC, and elsewhere). Originally heard on XM Radio, the program continued after the merger. To secure the partnership, Dylan's business manager visited the station in Washington, D.C., and met with XM CEO Hugh Panero. Dylan was said to be more interested in the musical aspect of the program, as opposed to the business aspect, and XM appeared to fit the bill.[28] Rebroadcasts of episodes were aired between 2011 and 2013, and they were later maintained as an online archive. The *Theme Time Radio Archive* website described the program as "a thematic journey through musical history" that includes "well-known and ultra-rare musical testimonies to the assorted concepts to form a thematic narrative through our collective consciousness."[29] Episodes were pre-recorded and included a combination of songs, stories, and fictitious or exaggerated phone calls and emails. Dylan's assumed authenticity imbued XM Radio with a sense of value, working to convince listeners that they would hear something exceptional on satellite radio. Listeners were meant to feel as though they had exclusive access to Dylan and his personal reflections on the songs he included in each episode.

Additionally, *TTRH* serves as a detailed compendium of Dylan's recorded musical output during his tenure as a radio host. As popular music scholar Keith Negus explains, Dylan's albums from the turn of the century, along with the radio program, "encapsulated [his] approach to his identity and music in the new millennium—reinvestigating the songs that inspired him in the 1940s and 1950s and returning to a mixture of the profound and frivolous, the comic and the caustic, that he explored so successfully during the first half of the 1960s."[30] The episodes, then, can be thought of as liner notes for fans that unofficially and informally credit the many musical and literary influences and collaborators that one hears on an album like *Modern Times* (2006). The radio program was essential listening for comprehending the range and scope of music and artists heard on the album. For example, in the song "Thunder on the Mountain," Dylan adapted lyrics from "Ma Rainey" by the influential blues artist Memphis Minnie. He changed key words, substituting "Alicia Keys" and "Hell's Kitchen" for "Ma Rainey" and "Georgia." The lack of formal credits on the album was a subject of debate in reviews as well as in academic writing on Dylan's larger body of work.[31] Some attributed the practice to a folk tradition of building on influences and predecessors and not treating authorship as it is dealt with in the realm of more commercial popular music. *Rolling Stone* music editor Joe Levy said of Dylan's *Love and Theft* (2001) that this was the "tradition he [came] from—blues and folk songs that have been endlessly handed down and adapted, stolen, mixed and remixed."[32] Dylan spoke to this practice in an article in the *Los Angeles Times* in 2004, explaining that "you can't just copy somebody. If you like someone's work, the important thing is to be exposed to everything that person has been exposed to."[33] Regardless of the valorization of the folk tradition that characterizes these quotes, Dylan's failure to credit his sources certainly merits criticism. He has a high level of star power and profits more generously from music than the artists from whom he borrowed at that time. But on the radio program, as explained below, Dylan would talk in detail about Memphis Minnie's influence on music history.

Rock and folk are shaped by a belief in authenticity, and this belief, for some radio listeners, sustains subscription through the notion that commercial broadcast radio is devoid of authenticity, while satellite radio enables a more unique listening experience. Folk has been distinguished by an ideology of anti-commercialism; it is a genre "considered as more authentic and real than the music produced by record labels."[34] Rock is also defined by its anti-commercial stance. Rock has drawn on attributes of intimacy and immediacy in its expression of authenticity.[35] This involves working to convince listeners of a lack of studio fabrication or of a strong connection to a notion of resistance or rebellion. While both genres are relatively fluid and are defined and policed just as much by fan communities, music journalism, and visual style as they are by their sound, the two collided and entered popular discourse in the mid-1960s with the advent of folk-rock, or electric folk. Dylan is regularly given credit for this hybrid genre due to his famous "plugged in" performance at the Newport Folk Festival in 1965. The performance became a storied moment in music history by which Dylan has been characterized as the sort of creative individual who can effectively transgress folk and bring elements of the folk genre, namely its political awareness, into the realm of rock. This hybrid genre would be an extension of the commercialism of an American folk revival beginning in the late 1950s that for some meant a lessening of the political content of the "earlier 'folksong movement'" but for others was marked by "a complex politics" of a "distinctly modern, technological society, wherein the LP format and the new medium of television were inextricable components of the cultural fabric."[36] The folk revival "was built on a contradiction," one that enabled Dylan to emerge as a folk music star, according to popular music scholar Lee Marshall: "It was a mass-mediated revival of a form of music that was against the mass media."[37] Dylan's perceived authenticity in the 2000s reflects his prominence within the popular histories of these genres. A prime example of this is his strategic use of words and sounds from the past, as indicated by the earlier "Thunder on the Mountain" example. Dylan and his radio voice advanced this mythologized history, and this became the means by which listeners were further convinced of the artist's authenticity and, in turn, the value of subscription satellite radio.

Dylan's star image embodies a notion of hybrid authenticity, a characteristic that also helps to make sense of the appeal of satellite radio.

He is perceived by many as being well-versed in music history and tradition but as also having his ear open to future trends and trajectories.[38] Both of these qualities lend themselves well to the archetype of the collector. This notion of hybrid authenticity is echoed in a *Modern Times*-era profile of Dylan written by critic Jon Pareles, in which Dylan's star image is characterized as straddling the past and the present: "[In the 1960s] Dylan transfigured pop songwriting with the shocks and disjunctions of modernism: ideas he found equally in the avant-garde and in old, weird folk songs."[39] Using *TTRH* as a platform, Dylan circulated his star image, one defined, in large part, by this notion of hybrid authenticity. There are intriguing connections between the hybrid authenticity of Dylan's star image and satellite radio. SiriusXM reflects a notion of hybridity because it has both a "broadcast" component as well as an expanding list of channels available only through its app; thus the satellite radio service is indicative of both radio's past and its future.

Dylan's career at the time of his XM radio show has also been described as a period of "memory work," one defined by *Modern Times*, *Time Out of Mind*, and *"Love and Theft"*, but also by the autobiographical book *Chronicles* (2004) and two major film projects: Martin Scorsese's documentary *No Direction Home* (2005) and Todd Haynes's *I'm Not There* (2006).[40] As an integral component of this period of memory work, Dylan's radio persona and radio voice worked to link the past with the present, and in the process can be said to have added insight into the various influences heard in his recorded music. Dylan's "reinvention of himself" as a DJ reinforced this sort of memory work because it enabled listeners to access "the intersection of Dylan's musical memories with the collective memory stored in the recorded archive."[41]

There are precise ways that *TTRH* amplified Dylan's hybrid authenticity, making effective use of radio to provide ancillary traces that complemented Dylan's recorded music of this period while simultaneously ascribing a notion of value and credibility to XM Radio. Three integral aspects of the radio program that collectively worked to transmit hybrid authenticity and a high level of synergy between the record and the radio are (1) Dylan's radio voice and its narration and curation of musical selections and their related historical contexts, (2) Dylan's privileging of lesser known yet highly influential artists through a supposed freeform radio format, and (3) a celebration of older technology to help craft an intimate radio voice to connect with listeners. These elements also reinforced ideas of authenticity in folk, in rock, and in Dylan's career more broadly speaking, especially against pop commercialism and commercial format radio.

First, when one listens to episodes of *TTRH*, Dylan's radio voice fosters a sense of intimacy and closeness with him as the celebrity behind the microphone. His vocal delivery and its rhythmic patterns are affective and add personality to the music he plays. Across the episodes, Dylan reaches into the past to select songs, many of which are not regularly heard on the radio. Further, he relies on the listener's assumptions about his musical expertise to convincingly comment on the wider cultural context surrounding his musical selections. The film critic A. O. Scott, writing about Todd Haynes's biopic, *I'm Not There*, says that it is only "a certain kind of artist [who] will comb through the old stuff that's lying around—the tall tales and questionable memories, the yellowing photographs and scratched records—looking for glimpses of a possible future."[42] This is the approach that Dylan takes with *TTRH*, playing obscure songs often from the 1940s and 1950s. Sounding experienced and weathered, Dylan's radio voice often mumbles and stumbles over words while simultaneously demonstrating knowledge of American music history.[43]

Dylan alludes to a broader context of social justice in episode 12, "Cars." He introduces Memphis blues guitarist Memphis Minnie and accentuates her place as an influential guitar player in the rock genre. As discussed previously, Memphis Minnie's work lives on in Dylan's "Thunder on the Mountain," although she is not credited on the album. On the radio program, however, Dylan highlights her significant contribution to rock and blues history, an important point to emphasize given the ongoing and unfortunate gender stereotypes that assume proficiency on the guitar to be a masculine domain:

> They don't make cars like they used to. A lot of things they don't even make anymore. And remember, there's a lot of things tomorrow that they're not making today. So get used to it. "Me and My Chauffeur Blues"; One of the great blues songs of all time; One of the great car songs of all time; One of the great chauffeur songs of all time; Sung by one of the great old ladies of all time. . . . Minnie began playing guitar in the late 20s and in all cases she was more than any man's equal.[44]

Dylan interrupts the song to say, "Memphis Minnie. She moved to Chicago in the 1930s, and added bass and drums. She was before her time. Anticipating the sound of 1950s Chicago blues." In this moment during the twelfth episode of the program, Dylan expresses a fondness for the high quality of vintage cars and uses this as a metaphor for Memphis Minnie's career. Technology plays a critical role in shaping Dylan's public persona via the music radio program and the artist's voice, both literally and figuratively.

The second way that Dylan distinguishes his program from commercial broadcast radio is by using a freeform radio approach to programming, which provides the radio host with a lot of control over the music played. Freeform radio programs "are as different as the personalities of DJ's, but they share a feeling of spontaneity, a tendency to play music that is not usually heard."[45] This programming philosophy is borrowed and used by Dylan to challenge the format logic of commercial radio. As with freeform radio, there is an emphasis on musical variety promoted by XM Radio. In *Chronicles*, Dylan writes that he often felt isolated while recording *Oh Mercy* (1989) with producer Daniel Lanois in New Orleans but found comfort in the city's infamous community-supported jazz, blues, and local radio station, WWOZ-FM. Dylan says that "WWOZ was the kind of station I used to listen to late at night growing up, and it brought me back to the trials of my youth and touched the spirit of it. Back then when something was wrong the radio could lay hands on you and you'd be all right."[46] WWOZ is a listener-supported radio station with a guiding musical aesthetic that works to mirror the musical activity of its home city of New Orleans, or at least certain aspects of that musical activity. In other words, it is not *entirely* freeform.[47] But a station like WWOZ shares an affinity with commercial freeform stations of the 1960s and 1970s, which were guided by a commitment to musical variety and diversity.

In episode 17, "Friends & Neighbors," Dylan reads a listener email, one potentially real but likely fictionalized. His response to the email communicates the unpredictability of his show. On *TTRH*, listeners will not hear songs routinely programmed by classic rock commercial radio. Dylan humorously prefaces this mandate by stating that the postal worker, a human being carrying and delivering physical media, is superior to their digital counterpart, email:

I like email but I miss the postman. I used to like it when he would come by and tell me what my neighbors were doin'. Oh well. Times change. Today's email is from Vernon Talbot. From Tampa, Florida. He writes, "Dear *Theme Time Radio Hour*, love the show. But I can't help but notice that most of the artists you play are somewhat obscure. What's up with that?" Well Vernon, that's a fair question. First of all, why should we play things you can hear anywhere else? On the other hand, the artists we play are interesting and deserve their moment in the sun.[48]

Dylan devotes his eighteenth episode to the theme of radio, playing a selection of songs that includes Van Morrison's "Caravan," Bonnie Owens's "My Hi-Fi to Cry By," and The Blasters' "Border Radio." During one of the episode's musical breaks, Dylan describes his relationship to the medium through a challenge he poses to the words of T. S. Eliot. Dylan's radio does not leave listeners lonely; it facilitates community, and we never know what we are going to hear:

> T. S. Eliot once said, "radio is a medium of entertainment, which permits millions of people to listen to the same joke at the same time, and yet remain lonesome." Well you're never lonesome when you're listening to *Theme Time Radio Hour*. Some radio programs play just one type of thing. But here, we're like New England weather. If you don't like what you're hearin,' stick around. It'll change in a minute.[49]

By emphasizing community and conversation, and through borrowing from a freeform radio format, *TTRH* neatly fits with XM Radio's promotion of musical variety across its channels. As with Dylan's star image and hybrid authenticity, sonic elements that seemingly refute standardization, predictability, and repetition circulate a perceived superiority over commercial format radio. This serves to accentuate XM Radio's perceived value and at the same time promote Dylan's recorded musical output of the era.

The third way that Dylan crafts distinction through programming is by incorporating the sound of old time radio (OTR) into his show. OTR is a fitting companion to the musical selections Dylan makes from decades past. The sound of OTR, those radio programs from when the medium was the primary form of entertainment, facilitates a feeling of community and nostalgia through Dylan's voice and the associated collection of sounds and songs heard on the program. A sonic teleporting to radio's past masks the more commercial aspects of Dylan's career and XM Radio. Further, the OTR aesthetic indicates a process of remediation, a concept that can also be applied to the previous sections on Dylan's radio voice and freeform radio. Scholars have curbed the rampant enthusiasm that frequently accompanies narratives of "new" media, often advanced by those companies profiting from new products and services. Instead, we are urged to acknowledge the various ways that new media carry forward key organizing principles and practices of older media. Shifts and changes in the digital era are in actuality less often about newness and more often about implementing established industry practices like licensing and branding.[50] Jay David Bolter and Richard Grusin define remediation as an instance wherein "new media are doing exactly what their predecessors have done: presenting themselves as refashioned and improved versions of other media."[51] This process involves a newer medium promoting a "claim of superiority" over the older medium it is refashioning.[52] In XM Radio's emphasis on its revolutionary qualities, a preoccupation with this discourse of superiority and excellence is evident. At the same time, its continuation of radio's essence as a medium that fosters feelings of community through private listening is heard on a program like *TTRH*.

In *Chronicles*, Dylan reflects back to a time when he first moved to New York: "I was always fishing for something on the radio. Just like trains and bells, it was part of the soundtrack of my life."[53] Throughout *TTRH*'s episodes, one hears sonic fragments from radio's past, such as radio jingles and vintage station promos. Over its long history, radio has facilitated feelings of community and connection through the act of listening.[54] According to a *Rolling Stone* review, almost every song on *Modern Times* "retraces the American journey from the country to the city, when folkways were giving way to modern times. The mood is America on the brink—of mechanization, of war, of domestic tranquility, of fulfilling its promise and of selling its dreams one by one for cash on the barrelhead."[55] Dylan uses themes of domesticity and stories about making connections through the airwaves in order to introduce and contextualize the songs he plays on *TTRH*. In episode 4, "Baseball," Dylan remembers listening to a baseball broadcast in his bedroom when he was a young boy. He reads a listener email that asks, "Dear *Theme Time*, I enjoy listening to the ball games late at night. My boyfriend says the radio keeps him up. What should I do?" In response, Dylan addresses his listeners to say:

> You should do what I used to do. When I was supposed to be asleep, I'd take the bedside radio and slip it under my pillow. Press your ear close to the pillow, which is what you're supposed to do with pillows anyway, and listen to the second game of the doubleheader without botherin' anybody else in the house. Millions upon millions used to do the same thing. Back when radio was king. And I hope you still do that with *Theme Time Radio Hour*. Your private pillow pal.[56]

Dylan's memory work of the era is sonically constructed via his on-air stories and the sounds of OTR that are mixed throughout the episodes. His preference for old media, communicated via a newer medium, indicates a notion of hybrid authenticity, one that shapes his radio voice and the intimate relationship he establishes with listeners through *TTRH*.

A program like *Theme Time Radio Hour* significantly contributed to the overall radio programming that has enabled XM Radio (and later SiriusXM) to compete for subscribers against commercial format radio and streaming services like Spotify. In return, *TTRH* served as a platform for Dylan to talk about music and its related historical and cultural context over 101 episodes, which sustained his relevance in a distinct but related field and promoted his recorded material during his tenure as a radio host. As with the genres of folk and rock, *TTRH* reinforced a notion of authenticity that concealed the more commercial aspects of radio and records. In some episodes, the car company Cadillac was mentioned as a sponsor. One entire episode, in fact, was organized around the theme "Cadillac." The company became fused with the music and thus acquired Dylan's stamp of approval. In 2007 Dylan also participated in a television advertisement for Cadillac that referenced his radio program. Clearly, there was a particular demographic in mind for XM Radio Dylan's radio program: middle- to upper-class folks who grew up with Dylan's music and now had significant disposable income.

Dylan's radio show exemplified handcrafted, exclusive radio for middle- to upper-class subscribers because it often sounded live and as though Dylan was engaged in unique conversations with the listener. This program helped the satellite radio service stand out against hostless playlists.

But this sense of liveness was carefully constructed. A *New York Times* profile of the show explained: "[Dylan] typically records from home or on tour, XM says, even though an announcer says the show is recorded in 'Studio B of the Abernathy Building,' to lend it a vintage aura."[57] Further, Dylan was not the only source of all the information and knowledge that listeners heard as songs were introduced and contextualized. Television writer Eddie Gorodetsky produced *TTRH*, and his music collection was often used for the show.[58] As one listens to the archived episodes, it is easy to be convinced that we are hearing music from Dylan's personal music library, but this is not entirely the case. The contradictions inherent in Dylan's hybrid authenticity can be applied to a music radio program on a "prestige" satellite service. Dylan's meticulous relationship to themes of time and space, and his effective use of media and technology, are of immense significance to his status as a musician, songwriter, and radio host.

Although Dylan's radio show ended in 2009, his presence remained in the channel lineup when Bruce Springsteen began to host his own show in the early months of the COVID pandemic in April 2020 (he also has his own 24/7 channel, E Street Radio). The program began not long after Springsteen released a memoir (*Born to Run*, 2016) and *Springsteen on Broadway* (2017). His SiriusXM radio show, *From His Home to Yours*, drew multiple comparisons to Dylan, a figure cited as a notable influence not only on the radio show but also on Springsteen's career. A review of the program's fourth episode said, "If you had any doubt that Bob Dylan is the father of Springsteen's country, an episode of *From His Home to Yours* still has not gone by without at least one Dylan tune."[59] There was an informal and accessible tone to Springsteen's approach as a radio host, not unlike Dylan's. It felt spontaneous because he was isolated by the pandemic and broadcasting from his farm. In the first episode, Springsteen says, "Hardest thing now, not being able to see, hug, kiss loved ones." Later in the program he says, "It's lonely down here on the farm. . . . Here's my good friend Bob Dylan, singing 'Beyond Here Lies Nothin.'"[60] In the second episode, the show's duration lengthened from the first episode's hour to an hour and forty minutes. Springsteen's celebrity status allowed for a substantial change to the overall program schedule.

Given the broader context of the ongoing pandemic and the political disarray leading up to the 2020 presidential contest between Donald

Trump and Joe Biden, the new radio program was discussed by journalists, and Springsteen himself, as having something to say about the state of America at that time. Radio, once again, became a medium to work through and think through the social and political issues of the day. Interviewing Springsteen for *The Atlantic*, David Brooks made reference to Woody Guthrie's "This Land Is Your Land" as part of a conversation about what America means and asked Springsteen, "We have an American national narrative that doesn't include everybody. So how do you think about the meaning of America, the American story?" Part of Springsteen's response cited the radio show: "But I think from here on in, the American musical conversation is going to be a cacophony of rap and pop and Latin music and so on, and there'll probably even be some room for an old guy and a little bit of rock music. On my radio show on SiriusXM, I try to include all those different voices."[61] Springsteen, like Dylan, was perceived as an authoritative collector, in terms of how he was framed in reference to the satellite radio program, and his assumed expertise became tied to ideas of unpredictability and exceptionality in music programming.

THE TASTEMAKER

While some radio hosts use their experience to navigate the past and select songs to play, others introduce subscribers to new music or lesser-known songs. SiriusXM has a channel branded as its own version of a college radio station that plays contemporary indie and alternative music, SiriusXMU (the shortened version of Extraterrestrial Music University).[62] On most weekdays, XMU airs *Blog Radio*, a program that runs from 8:00 p.m. to 10:00 p.m. eastern time. Each program features a different host who is affiliated with a music blog. The program is an example of how hosts and the related curatorial work of a music blog act as *tastemakers* within the overall channel lineup. *Blog Radio* is a site of convergence. It combines music blogs with satellite radio, and it's a site where music production overlaps with circulation and exhibition. SiriusXMU describes *Blog Radio* as where the "web's most influential music bloggers keep you ahead of the curve with two hours of handpicked music every day." Bloggers from *Carles.buzz* (formerly *Hipster Runoff*), *My Old Kentucky Blog*,

Brooklyn Vegan, Gorilla vs. Bear, and *Aquarium Drunkard* host shows that often coexist as playlists or mixtapes that are posted online. As of October 2022, *Carles.buzz* has been retired by its host, Carles, and there are now four *Blog Radio* shows each week.[63]

Before *Blog Radio*, XMU ran a weekly program called the *Student Exchange Program*. It aired on Sunday afternoons, and each episode profiled a different college radio station in the United States. The participating station would create the playlist, one that was intended to reflect the station's programming but also align with the regular indie music fare of XMU. In 2003 XM channel U-POP was also airing hour-long programs from college stations, thanks to a partnership between WorldSpace Satellite Radio and West Virginia University's U-92.[64] XMU was initially an XM station, and competitor Sirius featured something similar with *CMJ New Music Report*, which aired two hours of tracks from the top 20 CDs at college and non-commercial stations in the early 2000s.[65] On Jennifer Waits's college radio blog *Spinning Indie*, a number of *Student Exchange Program* playlists from 2008 are archived. The program began in April 2007 with a playlist from Emerson College's WERS in Boston. Stations were encouraged to include some selections from local bands, and on May 5, 2008, Colorado State's KCSU played thirty songs of which four were local selections by bands Slim Cessna's Auto Club, The Piggies, The Epilogues, and The Apples in Stereo.[66] It was relatively easy for stations to participate; the show's host, Josiah Lambert, advertised an email address for stations to apply to the program, and it then took a few months for them to air.[67] One music director for KSLU in St. Louis, Missouri, said that it was an exciting opportunity "to be able to bring our station from Internet-streaming to satellite radio" and to have that national exposure.[68] However, with the merger of XM and Sirius, staff cuts meant that XMU lost its program director and "Dean of Music," and the program was discontinued.[69]

Relying on a combination of the music blog, the radio, and aspects of podcasts, *Blog Radio* enables music bloggers to demonstrate their knowledge of up-and-coming songs and bands. Both podcasts and MP3 blogs are indicative of the convergence of platforms and interfaces and the listening practices that unfold in a multiplatform and multimedia context. Media convergence can be understood as a way to bridge old and new technologies, formats, and audiences.[70] The sort of amateur or do-it-yourself (DIY) aspect of music blogs and podcasts characterizes *Blog Radio*, on which a close connection to new music is emphasized, bypassing the more conventional promotional channels in the radio and record industries. A defining characteristic of podcasting is the prominent role of the podcast host as a curator and an authoritative voice that listeners come to rely on for news, information, or cultural content. Jonah Weiner, writing for *Slate*'s "Ten Years in Your Years: The Past, Present, and Future of Podcasts" series, connects podcasts and their hosts to "the essential promise of the Internet: a means for surprising, revealing, and above all enabling encounters with people, things, and ideas we didn't know."[71] The podcast host leads a process of discovery. This is made more powerful thanks to the voice, because voices are trustworthy and dependable, Weiner adds. In the digital age, when opinions and noise circulate with abundance and speed, podcasts can ground listeners in the comfort of the voice and the familiarity of one's own tastes and interests. This process of locating music for others also defines the MP3 blog. Elena Razlogova, in her work on freeform radio and MP3 blogs, describes music bloggers as people who "take advantage of cloud storage services to democratize music recommendations previously reserved for professional music critics. They curate MP3 collections of rare and non-Western genres.... There then follows a wonderful odyssey into hidden and often forgotten sonic worlds."[72]

Blog Radio lends an online materiality to SiriusXMU, one that is flexible and fluid. Other SiriusXM channels communicate materiality in similar ways. Recall, for instance, the discussion of vinyl records in shaping the Classic Vinyl experience. Throughout the 2010s, SiriusXM was increasingly visible in online spaces and developed special attributes of the service that were only accessible online or through the app. Time-shifting, archiving, and storing are all components of SiriusXMU's *Blog Radio* programs. *Aquarium Drunkard* has been diligent about sharing each program's playlist online, and often segments of the *Blog Radio* show can be downloaded as a mixtape. In the blog post that shared its February 6, 2015, playlist, a post on *Aquarium Drunkard* explained that Zach Cowie, also known as Turquoise Wisdom, was a guest during the second hour of the show. Listeners and readers could download the companion mixtape as an MP3 from the blog. Prior to the show's airing, the blog anticipated Cowie's guest host spot and informed readers of the unique perspective and expertise of the guest host. The playlist includes tracks by Soft Machine's Robert Wyatt and Alice Coltrane.[73] Artists like Wyatt and Coltrane deviate from the typical XMU playlist because the channel tends to play fairly recent releases by indie bands and artists.[74]

Blog Radio hosts do not simply use their online identities to extend the style and brand of their blogs into weekly two-hour audio programs; they also use their online identities to direct listeners to additional online sources where songs can be heard again or where extra songs can be found. The Texas-based music blog *Gorilla vs. Bear* regularly informs listeners about new tracks or exclusive content that has been shared online. On February 19, 2015, after playing a new song by Chromatics, "Just Like You," Chris, the host of *Gorilla vs. Bear Blog Radio*, told listeners that this was the first track to be released from the new album *Dear Johnny*, and that the song could be downloaded for free on Johnny Jewel's SoundCloud page (Johnny Jewel is a member of Chromatics, and as of 2024 the album was still unreleased). Later in the show, a track by Cash Wednesday was previewed, and the artist was introduced as "a brand new artist that we know nothing about and he or she just uploaded that track on SoundCloud last night." Chris said that he "was into it, so [he] played it on today's show and that's how we do things on *Gorilla vs. Bear Blog Radio*."

Along with *Blog Radio*, XMU has regularly aired special programs hosted by celebrity musicians. Most of these short, monthly programs have been hosted by men, like Patrick Carney of the Black Keys or actor Jason Schwartzman, which again advances the juncture of ideas about authenticity and expertise, and their shaping of prestige programming through rock-derived music genres and the technology of the satellite (something that can be said of the collector archetype as well). However, as of March 2022, Phoebe Bridgers has hosted a monthly show called *Saddest Factory Radio*, named after her Saddest Factory Records label. Descriptions of the show at its time of launch echo the purpose of the *Blog Radio* roster in terms of adding variety to XMU's regular programming, but in this case, instead of an influential music blogger, it's a well-known indie artist behind the music. SiriusXM said the show "will fall in line with the label's aesthetics and promises 'outside-of-the-box thinking and audio trust falls.' As host, Bridgers will guide listeners through artist-to-artist conversations and, of course, play her favorite songs."[75] Bridgers, like the *Blog Radio* hosts, enjoys a high level of creative autonomy and flexibility in programming music beyond the regular rotation of XMU, which is reflective of the tastemaker archetype. Perhaps in time these less-heard song selections will make their way into regular rotation, indicating the tastemaker's larger role in the overall channel lineup and in the cultural industries more broadly.

THE VETERAN

The final radio personality that makes its way into SiriusXM's music programming is the veteran (or at least the final one that is discussed in this chapter). Before discussing musician radio hosts who fit the veteran archetype, it's worth noting that radio hosts with storied broadcasting careers demonstrate key aspects of this archetype. "Cousin Brucie" Morrow is one example of such a radio host. Morrow joined Sirius fairly early in the company's history and left SiriusXM at age eighty-four in the summer of 2020. He joined Sirius in 2005 after New York's CBS-FM switched formats from oldies to Jack FM.[76] Morrow has a long history in radio. He's a popular host with a deep knowledge of music and with a carefully crafted relationship to listeners. A *Teen Life* feature from 1963, "Ask Cousin Brucie," hinted at his proximity to stars of the day, with readers asking him what Stevie Wonder looked like and how old he was.[77] In 1961 *New York Journal-American*, wrote, "When you're talking to Bruce Morrow, avoid words like 'disc jockey' and 'rock 'n' roll.' Substitute, instead, 'radio personality' and 'popular music,' respectively."[78] This characterization of Morrow as *more* than your average radio DJ carries through numerous trade press profiles from over the years. He left the radio station WABC to move to WNBC in the 1970s, and when asked about why he switched stations, Morrow said, "The station can't come up with the kind of money I require, so what's the point of negotiating? . . . My kind of radio seems to be long gone. Now, disc jockeys behave like computers."[79] The same profile said that a key aspect of his success was that he talked directly to listeners, never *at them*. Similarly, *The Soho Weekly News* said that meeting Bruce Morrow was "like having your transistor radio turn into a human."[80] He was described as being against the trend of stations becoming automated, wherein computers would select and play the music, play commercials, as well as close and open the announcer's microphone automatically.[81] In the 1980s, Morrow "sensed a radio revival" and began to do shows on WCBS-FM in New York; his *Cruisin' America* program was national in 1987.[82] Morrow's reputation as "human radio" was again referenced when he moved to Sirius.[83] He was fired from CBS-FM when it switched to a Jack format, which was rolling out in a number of North American cities, a format that relied very little on DJs, if at all.[84] In 2007 he signed a multiyear contract to remain with Sirius and would not return to WCBS-FM because it had shifted away from the music of the 1950s and 1960s, which he felt deserved "to have exposure, not be locked in a vault."[85] He was content to program the music of these decades with Sirius, a radio company with national reach.[86]

When Sirius partnered with the Rock and Roll Hall of Fame Museum in Cleveland, opening up studios in the museum space, Morrow hosted a special four-hour broadcast to celebrate the occasion in 2005. While on air, Morrow said, "I know I'm talking to tens and tens of thousands of new people. . . .You give me a little of your time, and you and I will become relatives very quickly." He described the music programming of the broadcast as skipping around the heritage of the Rock and Roll Hall of Fame.[87] Throughout this broadcast, Morrow not only solidified his place as a key figure of radio's sociability but also made apparent the prestigious programming of satellite radio, emphasizing heritage and history through the partnership with the museum.

Characteristics of the radio veteran are evident in the careers of many hosts on the SiriusXM roster, especially on channels like Classic Vinyl, where hosts program from, or have been celebrated by, the Rock and Roll Hall of Fame. For example, Dusty Street has been inducted into the Bay Area Radio Museum Hall of Fame and programs "Deep Classic Album Rock from inside the Rock and Roll Hall of Fame."[88] A 1985 interview with Street described her radio career in San Francisco and Los Angeles, stating that she had been "one of radio's most outspoken personalities."[89] Street was the first woman DJ on the West Coast, with a radio career dating back to the late 1960s, and the interview ended with her saying she could be making more money in another occupation but would rather be doing a radio show for young listeners, and that her "mind's more important than [her] bank account." Another host, Meg Griffin, was described by SiriusXM as a "bona-fide radio legend and one of a handful of DJs celebrated in the Rock and Roll Hall Of Fame and Museum."[90] A Boston-based music blog interviewed Griffin and explained that she had been a champion of freeform radio since her career began in 1975, known for playing punk and new wave on New York radio, and later, for having worked alongside Howard Stern.[91] Griffin worked at notable stations like KROK in New York and WMMR in Philadelphia before the "ever increasing corporatization of the radio business" in the late 1990s had her feeling as though the "liberties she'd grown accustomed to" were slipping away. She moved to college radio station WFUV, an NPR affiliate, taking a $70,000 pay cut,[92] then was courted by Sirius to host and serve as program director for a rock radio program that borrowed from the freeform format, The Loft.[93] As of fall 2022, Griffin was working on three different SiriusXM channels.

The veteran is not unlike the collector in terms of showcasing longevity as a pillar of the host's personality, but with the veteran there is less emphasis on one's extensive encyclopedic knowledge of music and more emphasis on their status as an anchor in a genre or style of music. In other words, listeners are meant to trust the host's long-standing connection to a band, format, era, or genre that is central to a particular channel on the lineup. As SiriusXM's music programming has expanded over the years, more veterans have been added to the channel lineup; they reflect a move to include a wider range of music genres for a wider range of subscribers. The radio hosts mentioned previously, who are not musicians—Morrow, Street, and Griffin—fit the veteran archetype but are without the same celebrity star power as the hosts discussed later. In comparison to the tastemaker, the veteran relies less on having an ear to the ground, and veteran figures are brought to the airwaves as foundational figures in their respective music genres. They tie a formative moment in music history to their careers after music.[94] A few examples of the veteran are LL Cool J, Lisa Loeb, and Marky Ramone.

LL Cool J's debut album is fittingly titled *Radio* (1985) and includes the single, "I Can't Live without My Radio." A review of his 2000 release *G.O.A.T* said that it had been "a century in rap years" between these albums, later referring to him as a veteran who was holding his own in the genre.⁹⁵ A 2018 press release by SiriusXM explained that LL Cool J's new channel, Rock the Bells Radio, would "allow listeners to look at, and listen to, classic Hip-Hop through the lens of current culture."⁹⁶ Over a sixty-day period in spring 2024, some of the most played artists included Slum Village, E-40, LL Cool J, and Warren G.⁹⁷ Before Rock the Bells Radio, classic or "golden age" hip hop was heard on BackSpin, a channel that went on hiatus in 2008 but returned due to subscriber demand.⁹⁸ Rock the Bells Radio brings a veteran presence to its slate of hip hop music channels, to accompany older or "classic" songs. It signifies the genre's centrality to the overall lineup and conveys a sense of nostalgia for hip hop listeners in a way similar to classic rock or the decades channels discussed in chapter 3. Rock the Bells Radio is named after "Rock the Bells," the first song on side B of *Radio*, which was released as a single in 1986. It explicitly conveys the medium of radio and the portable boombox, which is pictured on the album's cover. These older technologies, along with LL Cool J's veteran status, add a sense of legitimacy to the classic hip hop channel.

LL Cool J also exemplifies aspects of the collector archetype, namely through the use of his extensive career and legacy to curate and present classic hip hop music, as well as showcase "other innovators of hip-hop music."⁹⁹ Regular shows that air on the channel include *Furious 5 Playlist*, which involves guests sharing the music of their favorite MCs, and *Planet of the Tapes*, a weekly program that showcases the extensive classic hip hop music collections of hosts Geechie Dan and Diamond the Artist. LL Cool J's extensive career anchors multiple elements on Rock the Bells Radio, and these collector elements serve to add perceived value, or a notion of authenticity, to hip hop's place on the channel lineup.

Stay with Lisa Loeb debuted in May 2022 on the 90s on 9 channel, which also regularly features former MTV VJ Downtown Julie Brown as a radio host throughout its programming. Brown was the first host on the channel after a period during which it was hostless, after SiriusXM fired JoJo Morales and KT Harris to reduce spending.¹⁰⁰ Loeb's hour-long program airs weekdays at noon, 4:00 p.m., and 8:00 p.m. eastern time, combining stories about her personal life with music from the decade. A description of the show indicates that her celebrity appeal extends beyond music, into recipes, parenting, and stories about concerts she has attended, but also confirms her status as a standout artist from the decade:

"Lisa will also share stories about the music and events of the '90s very authoritatively because, well, she was there!"[101] Loeb's massively popular hit single, "Stay (I Missed You)" could be said to have helped define the decade of the 1990s, having been featured in the film *Reality Bites* (1994), which has a close association with the sound and image of Generation X.[102] A 2022 profile of Loeb emphasized her distinct personality and independence in the record industry in the 1990s, indicating a level of authenticity to her star text, one that serves SiriusXM well in its attempts to distinguish itself from commercial hit radio. The profile says, "She didn't flirt with the camera or sing about sex. She wasn't a part of a harmonizing girl group singing songs someone else wrote, often through the male gaze. She wore clothes that were, yes, unexpected but were her own decision." Loeb sang, wrote, and arranged her own songs "at a time when that was still considered rare." The profile also explained that Loeb in the 1990s was facing male gatekeepers in the radio industry who would only program one woman at a time on their stations: "'Oh, we already have a Sheryl Crow.'"[103] Loeb's veteran presence connects with Gen X listeners who are now the middle-aged and middle-class demographic that SiriusXM has continually targeted. Although Loeb's musical independence is a feature of the channel's brand, so too are wider lifestyle elements, those that are notably absent in many of the rock programs hosted by men.

Marky Ramone of the Ramones has a channel named after him as part of the online-only channels, Marky Ramone's Punk Rock Blitzkrieg, and a two-hour weekly program that is heard throughout the week on the 1st Wave channel. Ramone's veteran status is emphasized in descriptions for both the channel and the show, the former saying that he is the "long-time Ramones drummer" and the latter calling him the "legendary Ramone." He has played with the Ramones over two eras, he came of age during the legendary 1970s in New York, his band defined the counterculture of his time, and he has been inducted into the Rock and Roll Hall of Fame.[104] Ramone's radio show and channel play classic punk and new wave music, linking him to the formative years of punk during which the Ramones rose to prominence. Some of the most-played songs over a thirty-day period in fall 2022 were "Alternative User" by Stiff Little Fingers (eighty-eight times), "I Love Livin in the City" by Fear (eighty-seven times), and "The Legend of Pat Brown" by the Vandals (eighty times).[105]

A *New York Times* feature on Ramone in 2018, titled "How Marky Ramone, Punk Rocker, Spends His Sundays" mentioned that he went into the SiriusXM studios on 48th Street from his Brooklyn Heights apartment each week to record his show.[106] Ramone said that he was always compiling set lists for his radio show and for the 24/7 online channel.[107] The Ramones have been described as being "more popular and known [today] than they were at any other time in their actual career" in part thanks to songs being featured in TV and film soundtracks and commercials.[108] Another aspect of the veteran archetype, then, is ongoing relevance with a particular fan base and listener community.

Veteran hosts and their wider reputations gather music fans together within genre or taste communities. As LL Cool J said of his channel, "It's where fans can come and be immersed in the music, the culture, the energy and the history of Classic Hip-Hop. . . . This is the place to be if you want uncut, raw, pure, classic Hip-Hop."[109] Given the lengthy careers of these veteran hosts and their close connection to particular genres or decades of music, they are also active in other forms of cultural production. These radio-celebrity partnerships and their wider capacity for selling things, music and otherwise, have a longer history. Writing about advertising agencies producing "star-studded" programs like *Kraft Music Hall* in the 1930s and 1940s, Cynthia B. Meyers explains that major star performers like Bing Crosby helped to link celebrities to products and that radio could rely on stars as a sort of "'presold' talent—that is, talent already proven to attract audiences."[110] Having veteran stars on the radio to organize music programming around allows for an already established audience, or fan base, to gather as a listening community. At the same time, it enables the celebrity to promote their various endeavors and cultural work. LL Cool J is a film and television actor in addition to a musician. Loeb has worked in television, guest starring on shows like *Gossip Girl*, and has released children's music and books, winning a Grammy for Best Children's Album for *Feel What U Feel* (2016). Ramone performs with his band Marky Ramone's Blitzkrieg and is also an author, releasing a memoir in 2015 titled *Punk Rock Blitzkrieg: My Life as a Ramone*. Evidently these hosts use satellite radio as a platform to expand and extend their star texts across time and space and across cultural products, activating and appeasing a fan base that ideally will subscribe to SiriusXM if they do not already.

PILOTED LISTENING

The SiriusXM channel lineup features a mix of hosted and hostless channels and shows. Some hosted shows are live, in that the radio program is happening as it airs, and others are pre-recorded to air later but with a radio voice included. Using a host is a financial investment, and SiriusXM aims to determine where there is a need for, or expectation of, the radio voice, and which channels can be hostless. In the case of the 90s on 9 channel, a move to hostless meant the eventual return of the host, likely due to this being an unfavorable change in the minds of 90s on 9 listeners. In other cases, the first ten channels on SiriusXM's lineup are moving toward hostless, playlist-style programming, with key channels like 60s on 60 moving to channel 73 in November 2021 and becoming 60s Gold. It was replaced by the hostless The Coffee House channel, which upgraded to channel 6. It's tempting to think of this move on the channel lineup as a downgrade for the 60s channel. The first ten channels are a prime position. The 60s Gold channel now sits beside 40s Junction on 71 and 50s Gold on 72. The 40s and 50s channels were also once on channels 4 and 5, respectively. It's likely that the channel was moved higher on the lineup due to an aging audience, as well as the exit of key radio personality Cousin Brucie. Channels 7, 8, 9, 10, and 11 remain dedicated to decades, from the 1970s to 2010s, and channels 2 to 6 are devoted to youth-oriented hit programming, with SiriusXM Hits 1, Pandora Now, TikTok Radio, The Pulse, and The Coffee House. These shifting channels provide a sense of how certain demographics and, increasingly, younger audiences are prioritized across the first dozen channels, with older decades now mixed among niche channels like Siriusly Sinatra, B.B. King's Bluesville, and Elvis Radio. Certain shows are anchored by radio personalities, like the decades channels, while a channel like The Coffee House feels more like a playlist meant to be heard in the background, without a voice to break up the songs. It delivers a coffee shop ambience. In this sense, it is also cheaper to program without on-air talent, and these hostless shows are comparatively more akin to other music streaming platforms than their hosted counterparts.

For many satellite radio subscribers, the radio host remains essential. Otherwise, a subscription to SiriusXM largely emulates a subscription to any of the other hostless music services. The element of sociability still matters for a number of subscribers, and certain genres on the radio are enhanced by having an appealing radio personality to provide conversation and comments about the music played. Of course not all channels are hosted by celebrities, and many feature radio hosts with a wide variety of experiences, industry connections, and personalities. Rida Naser, for example, joined the Recording Academy's Class of 2022 and had prior experience creating a college radio station, operated radio events for 92.3 NOW/CBS Radio in New York, and moved from an internship with Hits 1 on SiriusXM to become the program director for channels BPM (dance/electronic) and The Pulse (Hot AC). Naser has also created the BPM Empowered festival, which programs women DJs, and she hosts a weekday four-hour show on BPM.[111] But in many cases, when the host is a celebrity or an authoritative music personality, as the archetypes of the collector, tastemaker, and veteran indicate, an influential human element guides the listening experience. These are personalities with authority who contextualize and introduce music and who converse with subscribers while on-air and online.

When considering the prominent role of the host, we can think about the total listening process, including interacting and conversing online in addition to listening to the programming, as a form of *piloted listening*. The host informs listeners why certain songs are being played and why they matter. They reify songs as important and crystallize their significance within styles of music and radio formats. And they debut new and unheard songs for loyal audiences. *Pilot* also refers to testing something new for the audience. I spoke with one host who was working live on-air out of the XM studios in 2016. When speaking about his role as a host, he emphasized that the on-air personality is absolutely influential in guiding the listener. The elements of "curation and the personality and timeliness that radio uniquely offers" are part of the larger music consumption process: "Okay, now go check out the rest of what this artist has to offer. Go to your local record shop and pick up the LP."[112] Listeners become "equipped with information" for the next steps, like buying a record or going to a live show. Not every channel on SiriusXM accomplishes this in the same way, and that's an intended part of the overall balance in the channel lineup.[113] There's also a "close relationship" between SiriusXM and other streaming services in terms of how listeners navigate the music environment: "Listeners are introduced to artists, new artists they might not have known otherwise with some context and some personality... and then [are] able to directly follow up on those artists in the wide ocean of streaming."[114] But what separates radio from most other competing services, according to this host, is "the compelling" combination of "personality, context, timeliness, and curation."[115] With a service like SiriusXM, the subscriber is also paying a monthly fee to have the radio personality do the work of guiding or piloting the listening experience.

Within the subscription satellite radio business model, the costly practice of acquiring celebrity talent has been balanced by music programming without hosts or with cheaper on-air talent. The move to acquire internet radio station Pandora may be indicative of a future in which hostless radio plays a more prominent role. Further, online-only channels and other internet radio services provided by SiriusXM encourage hostless programming. Such developments raise questions about how the company will continue to demonstrate value via music programming and whether or how music can help subscribers justify spending their subscription dollars in the streaming age. The next chapter examines the expansion of SiriusXM geographically, across the internet, and in programming, its celestial ambitions spreading beyond borders and to the clouds.

6 The Transnational, Technological, and Programming Expansion of SiriusXM

"Bringing SIRIUS XM to More Listeners on More Platforms than Ever Before," wrote CEO Mel Karmazin in SiriusXM's letter to its stockholders in 2010.[1] The goal of expansion was explicitly stated, and it would guide the newly merged company in years to come. When SiriusXM took on a $530 million loan from Liberty Media in 2009, the company was facing bankruptcy. Its reliance on automobiles had become a liability due to a decline in new car sales in 2008 and 2009 and the Great Recession in the US and beyond.[2] SiriusXM, as a public company, still faced pressure to grow. It has had to find new avenues for revenue and locate and secure new markets of subscribers. Following the merger, the company pursued subscribers across wider age and class demographics and debuted new channels both on the satellite and online, or app-based, services. Heading into the 2010s, competition from streaming music platforms like Spotify and Apple Music intensified, particularly as these services became available on the automobile dashboard. SiriusXM aimed to balance its "broadcast," or satellite, subscription service, which was still referred to in the trade press as the most central aspect of its business model, with new program offerings and services available on smartphones and laptops. Changes to SiriusXM's business model after the merger can be explained by the continued pressure to grow subscribers and revenue and to continuously construct perceived value for current and potential subscribers. But they can also be understood as part of radio's intrinsic adaptability. Radio is "an incomparably flexible, portable, and adaptable mode of mass communication."[3] Radio is a "'stealth' medium," says Michele Hilmes, "with a reach and an ability to evade control that makes its decades-long history of capture by the nation/state hard to believe as we 'think it differently' in the present."[4] Radio moves through the local, to the national, to the international and, simultaneously, the satellite radio industry and its business strategies and models shift from relying on a notion of outer space to inscribe perceived value in the service, to include, more emphatically, cyberspace and a subscriber's connection to music via the internet.

Exemplary of the expansion of music on SiriusXM is the international superstar Pitbull. On May 19, 2015, Globalization Radio, a channel based on Pitbull's music and 2014 album, *Globalization*, was launched by SiriusXM. The channel features contemporary dance hits from around the globe, mirroring the international reach of Pitbull's success, as well as one of his catchy nicknames, Mr. Worldwide. On channel 13, Globalization Radio occupies a prominent spot on the SiriusXM channel lineup and complements SiriusXM's outward-looking strategies to secure new subscribers, not only with respect to music from around the world but also across age and taste categories. Pitbull's wide-ranging reach and influence was referenced in a 2010 profile of his hit songs, where it was said that he has a "remarkable capacity to produce hits in all types of formats, genres and languages." In July of that year, he had a hit song on the *Billboard* Hot 100 and the Top 40 with Enrique Iglesias's "I Like It" and had three songs on the Latin Rhythm Songs chart ("Shut It Down," "Egoista," and "Alright"). These songs followed massive hits on the Hot 100 from the year prior: "I Know You Want Me (Calle Ocho)" (reaching number two) and "Hotel Room Service" (reaching number eight). Radio consultant José Santos said that Pitbull "is the real U.S. Hispanic" who attracts "Latin listeners and everybody else."[5] When Pitbull's channel launched on SiriusXM, he had his tenth *Billboard* top 10 hit with "Time of Our Lives," featuring Ne-Yo. In a preview of Globalization Radio, *Billboard* located the channel within Pitbull's "quest for world domination." In addition to music from Pitbull's catalog, the channel features music from his

Figure 18. Ad for "Pitbull Live at the Apollo" event in New York City. *Source:* SiriusXM press release, March 23, 2015.

"endless list of collaborations over the years" as well as live performances and DJ mix shows. The latter, like the airing of recorded live concerts on other artist-based channels, add another unique component to the channel lineup, and they are routinely heard on dance and hip hop channels. To launch the program, Pitbull performed at an exclusive, invite-only concert at the Apollo Theater in New York, an event that also livestreamed on the new channel.[6] Further, the channel anticipated SiriusXM's increasing presence across various mobile devices and online spaces. Alongside Globalization Radio came other Pitbull-fronted digital initiatives as part of a deal with "production giant" Endemol Shine North America. Two new original television series by Pitbull were introduced in 2015, with distribution across twenty platforms, including YouTube, Amazon Fire TV, and Roku.[7] Pitbull was the perfect artist to bring onboard to help soundtrack a wave of musical and geographic expansion.

This brief introduction to Globalization Radio offers insight into the theme of expansion, which has been recurring over SiriusXM's history. Before expanding online, Sirius and XM expanded into Canada. The satellite services were licensed in Canada in 2005 and later merged to form SiriusXM Canada in 2011. Developing satellite radio in Canada required the recasting of cultural policy for the circulation of Canadian music, namely in the form of Canadian content regulations for music. Expansion into Canada made effective use of an already existing North American satellite footprint, and the American company would, over the years, increase its economic control of the Canadian service. Expansion would later involve rolling out Sirius 2.0, which ushered in a new slate of channels, and 3.0, which included a redesigned app and a pronounced focus on on-demand programming and modes of customization for online listening. This chapter discusses the theme of expansion across geographic borders, media technologies, and the channel lineup (the next chapter critically considers the limits of expansion). In the post-merger years, SiriusXM aims to become much more than just a satellite radio company in competition with broadcast radio; it is moving to be an all-encompassing and ever-expanding music service that is tailored to the transformations in music listening brought forth by the platformization of culture. As the service moves beyond the automobile and onto mobile and personal devices, it maintains a sort of *cultural lifeline* to its subscribers, through which music programming and influential radio hosts spread across channels and platforms.

NATIONAL TO TRANSNATIONAL RADIO: SIRIUS AND XM COME TO CANADA

A theme of expansion is especially evident in the post-merger era, but before the growth of its streaming internet options, Sirius and XM expanded within the coverage of the satellite footprint by moving into Canada. The footprint spreads across national borders, but the satellite companies have controlled access to radio services within this space. In addition to expanding into Canada, SiriusXM was made available by terrestrial repeaters in Puerto Rico in 2009. The FCC had encouraged SiriusXM to expand the service to US territories "where technically feasible and economically reasonable" as a condition of approving the merger in 2008.[8] Although the satellite footprint already covered the island, the signal was obstructed by buildings and foliage, and the repeaters were needed to strengthen the signal. However, as chapter 1 explains, the footprint has been controlled in order to not unintentionally spill over national borders. In 2000 the US and Mexico reached an agreement that meant Mexico would not license any competing satellite services that would interfere with Sirius or XM along the border. Sirius and XM were not available in Mexico but used terrestrial repeaters along the border to complete their coverage.[9] A map of the SiriusXM footprint covers much of Mexico, although the service is not licensed to operate in the country.

As of 2005, satellite radio was licensed in Canada. With the satellite footprint covering Canada's major cities and populated areas, the infrastructure was in place to take advantage of potential subscribers in the nation to the north. Licensing satellite radio in Canada meant initiating policy discussions and debates about how to best implement an American service. Canada has a long history of cultural protectionism through cultural policy, and measures have been put in place to ensure that Canadian cultural products circulate via media like radio and television in order to help domestic cultural producers compete against the more dominant cultural industries in the US. Canada has so far decided to not implement cultural policy for the promotion of Canadian content for the internet. As far back as 1999, the Canadian Radio-television and Telecommunications Commission (CRTC) decided not to regulate new media, including internet services.[10] In order to license satellite services "for Canadians" a means by which to regulate Canadian content had to be established. As issues of cultural identity and sovereignty were cast in a Canadian and American context, one that accounted for the use of American satellites and the predominance of American programming, policymakers discussed how to best maintain foundational ideas about supporting Canadian culture for the new radio services.

In ongoing efforts to mitigate the influence of American media and culture, broadcasting policy in Canada has a tradition of defining its offerings as "for Canadians." As stated by the Broadcasting Act (1991), the Canadian broadcasting system should "serve to safeguard, enrich and strengthen the cultural, political, social, and economic fabric of Canada."[11] One of the ways that Canadian culture is reflected in Canadian broadcasting is through the use of Canadian content regulations. These requirements ensure that Canadian radio stations program a certain percentage of Canadian musical selections. These regulations were first implemented in 1971 as a condition of license for radio stations. Today, at least 35 percent of all popular music selections during the broadcast week on a commercial or community station must be Canadian and at least 50 percent must be in the public sector.[12]

Canada's national public broadcaster, the Canadian Broadcasting Corporation (CBC), was born from the idea that media in Canada should be for Canadian listeners. The 1929 Royal Commission on Radio Broadcasting stated that "Canadian radio listeners want Canadian Broadcasting."[13] This rhetoric has continuously permeated policymaking and public discourses surrounding the development of Canadian broadcasting. The American entertainment industry "constitutes a special case not found in any other country," producing "an enormous quantity of music, television programs, theatrical feature films, and Internet content, whose costs are largely amortized in the huge US domestic market that serves more than 300 million people."[14] American cultural products and services, "whose basic costs have been covered in the US market, can then be exported to other nations, including Canada, at low prices that cannot be matched by producers in these countries."[15] Considering global politics, economics, and the technologies that have facilitated signal transmission across borders, and given that it is largely an American model of media production, circulation, and regulation that is being exported throughout the world, the distinction between Canadian and American content can be lost on listeners. Thus, the Canadian government has played a role in encouraging Canadian culture and supporting institutions of cultural production.[16]

Although subscribers in Canada relied on American infrastructure for the delivery of satellite radio services, Canadians, as scholar Jody Berland writes, "pioneered the use of satellites to observe, map, and communicate with remote, frozen areas that were previously out of bounds for geological science and/or electronic media. Such experience helped to provide the technical and mythical infrastructure for nationhood; displayed on maps from above, Canada could thereby be 'pictured' as a modern nation."[17] However, when exploring the use of communications satellites in North America, the imbalances between the US and Canada in the production and distribution of cultural products and related infrastructures becomes apparent. In the wake of the Canada-U.S. Free Trade Agreement of 1989 and the North American Free Trade Agreement (NAFTA, signed in 1992 and coming into effect in 1994), the impact on satellite networks in Canada was said to be that US satellite carriers were not able to sell satellite capacity to Canadian customers or provide satellite-derived services in Canada using a US satellite. An American company, however, could establish a Canadian subsidiary to provide a service using a Canadian satellite.[18] Thus, the companies seeking the licenses would need to be Canadian, but they would deliver and operate the predominantly American satellite radio programming provided by American companies and satellites. The use of American satellites was ultimately allowed because no competing infrastructure was present in Canada, nor could it be easily and effectively developed; Canada was also without the necessary spectrum allocated from the International Telecommunication Union.[19] Regulators were concerned because if they didn't license and allow for the use of American satellites, Canadian consumers would find grey market access (the reselling of equipment or services of a company not licensed to operate in a national market), which had been an issue with direct broadcast satellite television.[20]

Canada's agenda of cultural protectionism has been notably evident in trade liberalization negotiations with the US. Cultural industries are exempted from NAFTA, and Canada kept "the right to review any investment (regardless of its amount) 'relating to Canada's cultural heritage or national identity.'"[21] Under NAFTA, Canada had a cultural exemption that allows for regulating any industry that affects Canadian culture. This includes telecommunications sectors like broadcasting.[22] The exemptions from international trade agreements allowed Canadian policymakers to apply cultural policy to the new satellite radio services for domestic operation. In late 2004 during a public hearing before the CRTC, three Canadian applications to provide subscription or pay radio services to be distributed by either satellite or digital transmitters were considered. The two satellite services to be offered by Canadian Satellite Radio Inc. (XM Canada) and SIRIUS Canada (Sirius Canada) were approved and implemented the following year, while the third, a digital service to be distributed via terrestrial transmitters and offered as a joint venture between Toronto's CHUM and Montreal's Astral, decided not to pursue operations. These Canadian services would be versions of their American predecessors, with non-Canadian-produced channels greatly outnumbering Canadian-produced channels. The Canadian services paid to use infrastructure and services from the American companies. For XM Canada, the American company would receive 15 percent of subscriber fees each month for basic service and 50 percent of net revenues for premium service subscriptions. XM also sold unused terrestrial repeaters to the Canadian company and agreed to provide technical assistance. XM owned 23 percent of XM Canada, with 11 percent voting interest.[23] Sirius also indicated that the company was reimbursed for costs incurred to provide Sirius Canada with the production and distribution of radios and information technology support costs. Sirius Canada paid a royalty based on a percentage of annual gross revenue for rights granted pursuant to the license and services agreement.[24]

In order to establish the service in Canada, Canadian content regulations would need to be sorted out for the satellite radio universe. Many musicians, policymakers, and music industry representatives in Canada were keen to allow the service to enter the country, while others were concerned with the lack of substantial measures of cultural protectionism. It was determined that the two satellite operators would offer at least eight Canadian-produced channels, each with at least 85 percent Canadian content. The program packages offered by the services could also have a maximum of nine American channels for every one Canadian channel, and the companies were given 150 days to accept these terms. Canadian Satellite Radio (CSR) was a Toronto-based company that paired John Bitove Jr. with XM Satellite Radio Holdings. Sirius Canada was a partnership between the CBC, Toronto-based Standard Broadcasting, and Sirius Satellite Radio. The CRTC stipulated that on the Canadian music channels, 25 percent would be new recordings by Canadian artists and 25 percent by emerging Canadian artists. A contribution of 5 percent of gross annual revenue would go to development programs for Canadian talent.[25] Some critics felt that relegating the Canadian music to a small selection of channels found amid the higher numbers on the channel lineup (as of this writing, in the 160s and 170s) was a disservice to Canadian content regulations, as they were easy to ignore or avoid, and that overall, the domestic music content featured on the new services would be less than what was heard on terrestrial radio.[26] But by the end of 2005 Sirius and XM had launched "up north" and that this meant "exposing acts to U.S. audiences." Sirius Canada began service on December 1 and XM on December 12. Sirius Canada had ten Canadian channels with four dedicated to music, which were available on the American service as well. XM had a total of eight Canadian channels with three dedicated to music. XM's Unsigned channel and Sirius's Iceberg were programming artists like Broken Social Scene, Metric, Luke Doucet, and Feist, who hadn't found much domestic airplay on commercial broadcast radio.[27] These channels were an outlet for Canadian artists to reach American listeners in the years before the rise of streaming music platforms.

CBC RADIO 3 AND THE CIRCULATION OF CANADIAN CONTENT

For advocates of the new services, satellite radio was touted as bringing new and emerging Canadian music into new spaces, namely the larger American market, but as also serving far and distant Canadian listeners who were underserved by terrestrial radio. Thus, the policymaking process for Canadian satellite radio drew upon the spatial advantages of satellite radio in comparison to the more limited range of terrestrial broadcasting. The CBC's Radio 3, a channel programming 100 percent independent

Canadian music, served as a beacon of hope for Canadian artists because its music programming drew from a wide variety of artists who had not yet received commercial radio play or achieved prominent levels of success, whether nationally or internationally.

The original idea for CBC Radio 3 was "a national FM network targeted at youth" that would be similar to the CBC's already existing Radio 1 and Radio 2. Due to the high costs of establishing the service on the FM dial, it was instead approved as a digital-only property based out of Vancouver, British Columbia. In 2002, Radio 3 rose to prominence as an online Adobe Flash magazine that featured an embedded streaming playlist specific to each issue. This magazine survived until 2005, when it was shut down due to budget constraints.[28] That same year, Radio 3 launched its first podcast. The podcast's music was generated from New Music Canada, one of the nation's largest music databases.[29] In 2006, CBC Radio 3 was Canada's most downloaded podcast, with at least 125,000 listeners per week.[30] Simultaneously, Radio 3 found a spot on the FM dial as a program featured on CBC Radio 2. The FM radio show played music from the Radio 3 website; it lasted until it was discontinued under a programming realignment in 2007.

Radio 3 helped the CBC become "an unlikely leader in digital technology."[31] The channel was well-equipped for satellite radio, and in December 2005, Radio 3 was available on Sirius. In a *Globe and Mail* article from November 2005, Alexandra Gill reflected on Radio 3's presence in Sirius's offerings, asking, "Is public broadcasting relevant in a 500-channel universe? Will Canadian-content quotas survive in the age of digital downloads, Web-based streaming and mobile music?"[32] The Sirius channel was programmed separately from the stream available on the CBC website until June 2009, when funding cuts resulted in the webstream becoming a simulcast of the satellite channel. Prior to the simulcast, the satellite station aired a mix of Canadian music (85 percent) and "international" music (15 percent), while the webstream was 100 percent Canadian. The amalgamation resulted in 100 percent Canadian content for both the webstream and the Sirius channel, with some exceptions for international music that had some Canadian character, such as a cover of a Canadian song. Radio 3 helped reconcile legitimate concerns that satellite radio was not sufficiently Canadian. The channel was a vehicle for taking Canadian music programming to listeners far and wide. In Andrea Warner's brief oral history of Radio 3, she writes, "Sure, everyone knows Canada's major export is lumber, but our biggest cultural export? Indie rock. And from Arcade Fire to Dan Mangan, it's been CBC Radio 3 that's provided the platform."[33] Warner connects the multifaceted service to Canadian culture, positioning independent music as a national export at the center of this relationship. The channel's high Canadian content and emerging artist focus were points used by satellite supporters to counter concerns that the new service did not showcase enough Canadian talent.

During the 2004 CRTC public hearing that took place in advance of the licensing of Sirius and XM in Canada, the president of Linus Entertainment, Geoff Kulawick, made a statement that presented the tensions at play in the licensing process with respect to how Canadian music would be featured. Kulawick said that Canadian artists did "get more plays on satellite radio in America than we received on terrestrial radio in Canada . . . but the plays we received were surrounded by tens of thousands of plays of other artists that weren't Canadian."[34] Moreover, the notion of "broad exposure" was questioned by Brian Chater, the president of the Canadian Independent Record Production Association (CIRPA). Chater argued that "the reality is that there is not broad exposure by listeners, as it is, by the applicant's own admission, a niche subscription service with a limited market." Chater claimed that CIRPA was also perplexed by the CBC's involvement with the international radio service, particularly its proposed $13.4 million contribution to Sirius's first two to three years of licensing. "CIRPA has diligently searched the Broadcasting Act," said Chater, "and nowhere can it find wording that says [the CBC is mandated] to be a distributor of foreign program services or that it should be programming into the United States[,] unless we have missed something in our reading."[35] In contrast, Neil Dixon, the president of Canadian Music Week, argued that "one of the most difficult things we had to do in promoting independent music on an independent label was getting it outside this country."[36] Dixon championed the advantages of satellite radio in comparison to terrestrial radio, carefully keeping Canadian content and cultural identity in focus: "So, as much as we appreciate the fact that the 35% Canadian Content is there that gets artists started and boost[s] their career[s] [through commercial radio], the opportunities that satellite radio offer are sort of unprecedented for Canadian artists."[37] Gregg Terrence from Indie Pool, an organization that aggregates and distributes music by and for Canadian independent artists, mentioned a divide between larger organizations like CIRPA and smaller organizations such as Indie Pool. He claimed that the CIRPAs "of the world" do not "represent Canadian independent recording artists." CIRPA, according to Terrence, represents the second tier of the music industry, that of the "major indies." Tier 1 is the "major label artists, the ones that everybody knows," and Tier 3 "is the other 95% of the music industry.... That is the 24,000 artists in Canada that are independent." The third tier, according to Terrence, is what he represented. Terrence's clients and artists "overwhelmingly support Canadian Satellite Radio, for one, because of the Canadian talent dollars that they are investing in Tier 3."[38] Terrence added that hundreds of Indie Pool's eighteen thousand artists were getting airplay on XM radio, whereas only about five were receiving airplay on Canadian commercial radio.

Some artists from genres and musical styles that can at best be heard on the margins of commercial radio programming also spoke in favor of licensing satellite radio. Singer Susan Aglukark said that she believed that satellite radio would finally help to give Indigenous artists "a voice that they so richly deserve." She said that they would "finally reach people in all areas of our country, including those listeners that have had far too little choice for far too long." Aglukark suggested that there was room for expanding the types of music heard on Canadian radio, and she anticipated that a more diverse and varied Canadian cultural identity might be heard on satellite radio. The late Jeff Healey added that as a musicologist, a musician, and a manager of a music venue, he was "very interested in seeing Canadian talent and the awareness of it spread around the world as much as possible," particularly in the "specialty fields" of jazz, traditional jazz, and blues.[39] A number of the country's independent artists were optimistic about the proposed satellite services, while those who represented more established Canadian artists, those who had already found success as Canadian content staples on commercial FM radio, were more apprehensive about the service or, perhaps, were wary of new competition for commercial broadcast radio. A channel that programmed emerging and independent Canadian music, such as Radio 3, and the artists it played appeared to have much to gain from the satellite services.

After the public hearing, the CRTC outlined its licensing framework for Canadian satellite subscription radio undertakings. The commission explained that there was no clear indication that a Canadian-owned and -operated satellite subscription service would be feasible for the foreseeable future, but that the use of American-owned satellites could introduce satellite subscription radio into Canada in a way that would contribute "substantially to the fulfilment" of the objectives of the Broadcasting Act.[40] The CRTC claimed satellite radio would increase the diversity and comprehensiveness of programming choices available to Canadians, and it would cover the entire country, bringing new services to rural and remote areas where radio options were limited. Moreover, the services were said to increase monetary support for Canadian musicians, particularly new and emerging artists, and would offer Canadian radio listeners multi-channel services in their cars and other vehicles using a proven technology.[41] By turning to the space of the footprint, proponents of satellite radio could recite radio narratives of connecting disparate listeners and sharing in the experience of a wider variety of Canadian music and culture. Strategically, this also helped to mitigate legitimate concerns that Canadian music would be lost in the expansive satellite services.

Sirius Canada and XM Canada merged in 2011, and in 2012 SiriusXM Canada renewed its license with the CRTC. During the license renewal, a shift in ideas about the delivery of satellite radio was apparent. Namely, the comparison between satellite radio and terrestrial radio transitioned to one between satellite radio and other mobile listening devices and services. As with the merger in the US, proponents of SiriusXM Canada argued that satellite radio was not just in competition with broadcast radio but increasingly with iPods, internet radio, and media platforms.[42] A shift in the spatial understanding of satellite radio, within both policymaking and the delivery of programming, moved regulators to consider new relationships between users and programming. This shift involved moving from thinking about satellite radio operations as being predominantly determined by a satellite's footprint to considering mobile and personal devices, including, but no longer limited to, the automobile. A spatial preoccupation with the footprint, which had juxtaposed Canadian cultural sovereignty with the potential for Canadian artists to penetrate American markets in 2005 when the services were first licensed in Canada, changed to reflect new issues and questions tied to mobility and individualism when the services merged in Canada in 2011 and when the license was renewed the following year.

In the policy decision outlining the Canadian merger, the CRTC claimed that "regulated and unregulated competition, combined with the investment requirements of the [satellite radio] business, have proven challenging to their business models."[43] Here, the satellite services were not considered isolated entities in a specific market, but rather one method of audio delivery among many. The changing technological landscape anticipated during the American merger was evident during the Canadian merger, some three years later, where "the potential for significant public detriment stemming from the proposed merger" was thought to be mitigated by the increasing availability of other audio services.[44] The range of competitive devices and communication outlets had increased from broadcast radio, iPods, and internet radio (with hints of podcasts and mobile phones) in 2007, when the American companies pursued a merger, to reflect the numerous personal computing and entertainment devices that any one listener might own and use on a regular basis in 2011, and most significantly, the increased power and sophistication of the smartphone. Further, the satellite services committed to offering subscribers the ability to access content on the existing online media player. The CRTC considered that this commitment, among others, offered "clear benefits" to broadcasting and to consumers in Canada.[45]

In the 2012 license renewal for SiriusXM Canada, the requirement for a minimum of eight Canadian channels was reworded to state that no less than 10 percent of the total number of unique channels had to consist of original Canadian-produced channels.[46] All Canadian-produced channels were to ensure that a minimum of 85 percent of all musical selections during the broadcast week were Canadian, and at least 25 percent must feature musical selections that had not reached a position on any of the major "hit" charts.[47] During the public hearing that preceded the license renewal, John Bitove, the chairman of SiriusXM Canada, noted that Canadian satellite radio had accumulated 2.1 million subscribers. Successes from the licensee's first term were emphasized by Bitove, such as serving "parts of Canada others just don't find profitable enough to serve" and providing "Canadian artists with a new platform for their talents to be displayed in Canada and across North America."[48] However, anxieties surrounding the changing technological landscape quickly made their way to the forefront of the hearing. The president and CEO of Sirius XM Canada, Mark Redmond, noted that the company was "in a dogfight for what you might call 'control of the dashboard.'" He explained that the company had invested heavily in getting vehicles equipped with satellite radio and had subsidized this cost, and now it faced competition from internet and apps-based, in-vehicle audio options.[49] Redmond claimed that the company preferred that people listen through the broadcast network, but if subscribers "want to ultimately listen on their iPhone . . . we should give them the ability to do that. If they want to stream it at their home on their computer, we should give them the ability to do that because at the end of the day, they're going to be more attached to our content and a better subscriber for us on the long run."[50] Redmond's comments signified a realignment, in which the consumer now required an attachment to the service and access through a variety of sources and devices. Automobile access was still the primary means for listening to satellite radio, but new components and aspects must be integrated into the service. Satellite radio became aligned with a variety of spaces. Those referenced by Redmond included the home, the cottage, the boat, and the secondary car—those spaces aligned with financial success and individualism. And this spatial realignment was driven home by Bitove when he said, "I thought we were much closer to a radio business when we applied seven years ago, but we are so drastically different."[51]

After the 2012 license renewal, SiriusXM Canada would become *more* American by virtue of its ownership. In May 2016 Sirius XM Radio Inc., the parent company of Sirius XM Canada Holdings Inc., demanded a payment of US$33.9 million from activation fees over the ten-year license agreement. At this time, the American parent company had a 32 percent stake in the Canadian company, with the CBC owning around 12.5 percent. Some analysts felt that this request for payment followed rumors of a potential buyout of the Canadian company and a move to take the Canadian company private.[52] Merely a week or so after this news was reported, SiriusXM was said to be increasing its economic interest in the Canadian operations to 70 percent. SiriusXM and Sirius XM Radio Inc. entered into agreements with Sirius Canada Holdings

Inc. to recapitalize SiriusXM Canada. SiriusXM and certain shareholders in Canada were going to acquire shares of the Canadian company that were not yet owned. SiriusXM Canada would remain under Canadian voting control, with Slaight Communications and Obelysk Media combining to own 67 percent of voting shares and 30 percent of economic interest. SiriusXM now had economic ownership of 70 percent of the company and 33 percent of the voting shares. This also meant that the CBC would cease being a shareholder in SiriusXM Canada. Along with this recapitalization, SiriusXM Canada was no longer a publicly traded stock.[53] According to one article by The Canadian Press in *Marketing*, this "going-private plan would allow a U.S. company to move in."[54] SiriusXM's annual report for the year 2017 announced that as part of a services agreement, SiriusXM Canada would pay 25 percent of its gross revenues on a monthly basis to SiriusXM until the end of 2021 and then 30 percent each month afterward (an advisory services agreement would also have SiriusXM Canada pay 5 percent of gross revenues on a monthly basis). At the time of this recapitalization, SiriusXM Canada had 2.8 million subscribers.[55] A few years after the recapitalization of SiriusXM Canada, the CBC would no longer provide music programming to SiriusXM. In October 2022, it was announced that Radio 3 was to be removed from the channel lineup. CBC Radio 3 (which had changed its name to Radio 3 Classic, channel 162) and CBC Country (channel 171) were cut, and the latter was replaced with a Canadian country channel programmed by SiriusXM. The French-language CBC music channels, ICI Musique Franco-Country (166) and ICI Musique Chansons (163), were also removed.[56] This decision sparked an uproar among artists in Canada and supporters of the nation's music industries, namely in smaller, more independent music circles. SiriusXM had become a major income source for Canadian artists, particularly by comparison to the low royalty payments from Canadian commercial radio and streaming platforms. One headline in the *Toronto Star* read, "'Final Nail in the Coffin': Why SiriusXM Dropping CBC Radio 3 Is 'Potentially Catastrophic' for Canadian Artists."[57] SiriusXM benefited many Canadian artists and labels because of the lucrative nature of satellite radio royalty payments. For artists, a royalty payment could be about $50 per play,[58] divided between the artist and the owner of the song's master (typically the record label). For independent artists without superstardom and massive streaming success, airplay on Radio 3 was especially lucrative.

However, even with the removal of the CBC channels, SiriusXM still had to ensure that no less than 10 percent of channels were Canadian, and 70 percent of the programming on these channels must be Canadian. Further, 25 percent of music must be new (released within the last six months), and 40 percent had to come from emerging artists.[59] *Emerging* means that the artist hasn't charted or reached a Top 40 position on pop/ *Billboard* charts or Top 25 on country charts.[60] This status lasts for thirty-six months after an artist reaches these chart positions. For its 2020 license renewal, SiriusXM Canada attempted to change the duration of "new" status to eighteen months (up from six) and emerging status to "not yet achieved Gold record sales" or not reached Canada Top 200 *Billboard* chart or Top 20 on Nielsen. The company also tried to change emerging status to a duration of forty-eight months (up from thirty-six).[61] Arguably, these attempts aimed to circumvent the programming of new Canadian music that was without hit status. These proposals were denied because the CRTC could not simply change these definitions, which had wider implications for other cultural institutions, but the commission said it would revisit them. Such an attempt by SiriusXM once again reflects the sort of power that a major media company wields in policymaking.

With the removal of the CBC music channels, SiriusXM introduced Mixtape North, a channel devoted to Canadian hip hop and R&B. The satellite service would still run the Indigiverse, a Canadian-produced channel that played Indigenous artists, as well as North Americana, an online channel that played "yesterday's folk, rock & roots." SiriusXM also programs The Verge, a channel that SiriusXM claims "continues to offer the best in new and emerging Canadian indie and alternative music, including many of the artists heard on CBC Radio 3." But some artists have said that channels like The Verge are "guided by corporate editorial interests, and are more likely to resemble mainstream commercial radio."[62] Certainly the new channels are without the public media mandate that shaped Radio 3, which aimed to reflect regionality and diversity in Canada. The channel description for Mixtape North also mentions massively successful commercial artists like Drake, the Weeknd, and Tory Lanez. Since the channel is still new at the time of writing, it remains to be determined whether or not it programs widely from lesser-known and/or independent artists. The most frequently played artists over a month in early 2023 included Drake, The Weeknd, and DJ Khaled, but also less mainstream Canadian hip hop artists like Snotty Nose Rez Kids, TOBi, and Aqyila.[63] Rez Kids are independent, while TOBi and Aqyila are on major labels. Most likely the move to remove CBC channels is due to SiriusXM wanting more control over programming in the more regulated space of Canada (by comparison to the US). The loss of channels with a public media mandate is a blow to music in Canada, no doubt, but this is a massive commercial subscription radio company with a long history of working to alter cultural policy in its favor, and a move like this comes as no surprise. These changes also mirror the broader strategy implemented under the American merger, with program redundancies and outside programming and partnerships coming under SiriusXM control.

MOBILE LISTENING, CULTURAL LIFELINES, AND SIRIUSXM 3.0

After moving into Canada, the satellite services increasingly became mobile and available online during the 2010s. The company was committed to finding new subscribers and adapted to changes in technologies and listening practices. With the use of streaming music platforms on the rise, SiriusXM developed an app for listening to streams of its channels; over time, the company has worked to improve its functionality. Steve Blatter, the senior vice president and general manager of music programming at SiriusXM, said in 2015, "You don't come across the college kid that aspires to be a radio DJ as much as you used to. Why would you want to be a radio jock? You can go on YouTube and build a following."[64] During an interview with one SiriusXM on-air personality, I was told that there has been a shift in which more subscribers are not receiving the satellite feed but are instead streaming on computers and on the smartphone app. SiriusXM has become "a much broader sort of content company" with on-demand capability and social media components.[65]

The drive to continually increase subscriptions also means looking beyond luxury vehicles and beyond listeners in a higher income bracket.

However, the company still emphasizes its exclusive and premium programming offerings. In late 2010, as US car sales were rebounding after the recession of 2008 to 2009, it was reported that "satellite and Internet radio services [were] making big plays for drive time." In an effort to branch out from its earlier focus on upscale automobile models, SiriusXM brought the service to a wider range of vehicles, with the expectation that it would be installed in 60 percent of all new light vehicles sold in 2010, up from 21 percent.[66] SiriusXM was hoping to outperform competing services like Rhapsody, which in 2013 offered stations that "mimic[ked] SiriusXM offerings" and catered to the vehicle's "lean back" listening experience.[67] Apple also partnered with automobile companies like BMW and Toyota.[68] In 2015, SiriusXM reiterated that "programming is the foundation of our success, and our unique and superior content is what continues to attract new subscribers and retain our existing ones."[69] At the same time, the company acknowledged that "efforts to increase the penetration of satellite radios in new, lower priced vehicle lines may result in the growth of economy-minded subscribers; our work to acquire subscribers purchasing or leasing pre-owned vehicles may attract subscribers of more limited economic means."[70] The company explained that these tactics might negatively affect profitability, as they might increase subscriber churn (turnover) or might not match the prices and margins consistent with previous subscriber-listeners. Seemingly, the company was uncertain whether targeting lower income listeners would pay off.

New online features have been debuted during successive waves of expansion by SiriusXM, dubbed SiriusXM 2.0 and SiriusXM 3.0. While SiriusXM 2.0 involved the launch of new channels thanks to an increase in bandwidth, SiriusXM 3.0 included an enhanced user experience on the app and web-based streams. As part of the 2011 debut of SiriusXM 2.0, a suite of commercial-free Latin music channels was introduced. The release of these new channels followed years of SiriusXM targeting Spanish-speaking listeners.[71] These channels were available online, indicating that the online-only channels were becoming a key growth area for the service, ideally driving subscribers to purchase packages that would grant access to both tiers. New music channels included Caliente, Viva, and La Mezcla, which paired Latin music stars with pop stars like Lady Gaga and Britney Spears, who were said to be popular among

Spanish-language listeners.[72] In August 2012 it was announced that the first artist-hosted show for bilingual listeners had been created. The Colombian singer/songwriter Juanes would host *Yo Soy Juanes*, a monthly program to showcase his favorite music, "from traditional Colombian folk songs to Tony Bennett."[73] The growth in Latin music on SiriusXM was framed as part of the ongoing focus on audiences that were left out of commercial broadcast radio in America. Tomas Cookman, the president of alternative-Latin record label Nacional Records, said, "The Pandoras and the Siriuses are the ones giving revenue [to us]."[74] SiriusXM 2.0 marked an expansion through increased bandwidth for the service as well as an expansion into demographics that had not been previously prioritized as (potential) subscribers.

Version 3.0 of the app allowed users to exercise more control over how they listened to songs and individual shows. It built on the MySXM feature that was released in 2013 and grew the service's on-demand programs. In April 2013 *Billboard* described MySXM as SiriusXM's "answer to Pandora." This feature created "an interactive Internet radio service by allowing users to personalize existing SiriusXM stations." MySXM was "powered" by Omnifone and the Echo Nest, and it used apps created by QuickPlay Media. It allowed users to alter more than fifty online channels, initially, with plans to expand. These select channels came with a set of slider bars to make variations in the music, such as playing older artists or more recent artists, faster songs or slower songs.[75] For example, The Coffee House online channel (a version of its "broadcast" counterpart) allowed users to turn the dial from "Depth" to "Familiar," from "Older" to "More Recent," and from "Coffee Covers" to "Originals." For the 40s on 4 channel, users could move between "Supporting Acts/Headliners," "Vocals/Instrumentals," "and "Slower/Faster." The introduction of this feature was characterized as part of a broader move by music and radio companies to "capture the social, interactive nature of the Internet."[76] A new, redesigned, and re-engineered internet radio streaming platform with on-demand functions as well as "enhanced programming discovery and the ability to connect with content currently playing across . . . commercial-free music, sports, comedy, news, talk and entertainment channels" was slated to begin operation in the first quarter of 2015 as part of the 3.0 rollout.[77] New developments in the delivery of satellite radio, such as MySXM and its on-demand features, illustrate the increasing importance of online features to the satellite radio industry.

Prior to debuting Sirius 2.0 and 3.0, SiriusXM made efforts to boost the mobility of its service beyond the automobile. Sound studies scholar Michael Bull argues that mobile music devices like the Apple iPod represent "a Western narrative of increasing mobility and privatisation."[78] While the iPod is its own unique device, Sirius and XM had experimented with a number of personal audio devices over the years. As early as 2005, Sirius and XM were eyeing home electronics and portable devices as growth areas that were outside of the automobile.[79] Examples of products that helped facilitate beyond-the-vehicle listening were the Stiletto 2 wireless satellite radio receiver by Sirius, which had time-shifting capabilities;[80] the Squeezebox Duet, which included a wide range of streaming internet-based music sources, including eighty channels of Sirius Satellite Radio;[81] radio apps on mobile devices like BlackBerry's Pearl and the Curve, as well as the iPhone, which were all announced in 2008 to 2009; and, the XM SkyDock, which enabled those with an internet subscription package to turn their iPhone into a satellite radio receiver through the cigarette lighter of a vehicle and the FM transmitter.[82] Debuting in 2011 was the Lynx portable radio, which allowed listeners to pause, rewind, replay, and store programming.[83] Moving to compete with mobile listening devices, satellite radio became increasingly expansive and tied to individual experiences of sound and music available via phones and mobile listening devices. Evidently the smartphone has won out against these other products, which is why the SXM app has become the gateway to streaming satellite radio and accessing online channels.

With satellite radio becoming increasingly mobile, programming extends into online spaces, providing listeners with new platforms and avenues to interact with DJs, music, and, to use a term unfortunately commonly overused in the streaming era, content. We might think of these mobile, individual, and personalized spaces and points of musical contact as *cultural lifelines*. This notion highlights the continuous connections, or streams, between users and content through a variety of mobile devices and services. Radio personalities are able to maintain relationships with listeners beyond their on-air responsibilities and into the sphere of social media and mobile listening. One radio host explained that the biggest personal benefit to being on-air is connecting with listeners on social media, playing new music, "and being able to follow artists together and reacting to what's happening on the channel and continuing the conversation online." The connection extends beyond music, and the host can "hear what people think about the world and about politics and about life, much broader than the music that [is played] on the channel."[84]

In both the previous chapter and this one, the crucial role of radio hosts is emphasized. SiriusXM, in a post-merger context, still champions the more human elements of its radio service while also expanding its online services. As such, it both embodies aspects of music platforms and maintains features of radio's past. In their book on platforms and cultural production, Thomas Poell, David Nieborg, and Brooke Erin Duffy explain that platformization involves the movement of digital platforms and their economic, infrastructural, and governmental extensions into the cultural industries, as well as the larger ways that platforms organize labor and practice within and around these platforms.[85] In *The Platform Society*, José van Dijck, Thomas Poell, and Martijn de Waal explain that selection, datafication, and commodification are key tenets of platformization;[86] Tiziano Bonini and Alessandro Gandini argue that music selection and curation are key to understanding the power of platforms.[87] Further, in Jeremy Morris's writing on Spotify and platform effects, he centers the role of optimization in enabling music discovery in a "sea of content."[88] In many ways, through expanding music programming and guiding listeners through satellite music streams, SiriusXM has both paralleled and foreshadowed the platformization of music listening.

A cultural lifeline maintains the spatial essence of the footprint, in that satellite radio still offers access to content within the footprint, but it also emphasizes the consumer's desire to easily and readily access content from any number of devices from a mobile phone to a receiver in an automobile. Sociology and communications scholar Paolo Magaudda traces the connections between music platforms and smartphones. Drawing on user experiences, Magaudda demonstrates how listening is shaped by access to reliable internet.[89] Thus, the notion of a lifeline accounts for the anxiety or fear of not being able to access content whenever, wherever. The SiriusXM host I spoke with said that "most people who subscribe and listen all the time would very much regret having that pipeline of content and personality turned off at any point." Lifelines also reflect a user's dependence on mobile media for a number of reasons, such as the development of smartphone apps as monitors of health. A *New York Times* article on this topic cites a doctor who jokes that many of his patients are "surgically connected" to their smartphones and describes devices that measure blood pressure and chart heart activity.[90] Mobile devices with internet connections are increasingly vital avenues for accessing information and culture. And while connections to music are not as vital as connections to statistics that indicate one's health, music is intimately tied to one's sense of self and identity.

As with the space of the footprint, lifelines are shaped by their own specific power relations and political and cultural issues. Lifelines summon mobility politics and questions of access. Who, for instance, can afford to maintain subscription fees and monthly mobile phone bills and participate in a networked society? Mobile devices range from simple and affordable to expensive, multifunctional devices, and thus the transformation of social space to consumerist space differs on a class basis; the shift is not universal.[91] Utopian ideas of "hypermobility" and "instantaneous communication" motivate business strategies and government policies.[92] How, then, does content converge across devices, and how do content providers ensure access from, and within, a variety of spaces and places? With a pronounced focus on mobility and constant connectivity comes a pronounced use of listener data in programming decisions. This also raises concerns with respect to privacy and surveillance when it comes to how listener data is used and integrated into business strategies, particularly in a wider context of platform capitalism.[93] A SiriusXM host said that with the earlier satellite radio model, it was difficult to understand the listener base because the feed was moving from the satellite into the car with nothing going back "up from your car into space." Now, with listeners using the app or listening online, there are more metrics and data that reveal what people are listening to and when.[94] An online privacy policy from 2015 explained that the company collected information when a user subscribed and used the service through the site, media player, or mobile application, and also through the use of social media applications linked to these services.[95] The policy informed listeners that they might also receive information from the media player, which had a personalization feature, allowing

SiriusXM to collect anonymized information including the channels they tuned to, the shows and the music they listened to, and the items they chose to download and listen to from the catalog of on-demand shows at a later time. A subscriber's listening orbit could be tracked and integrated into the shaping of future program offerings.

The notion of lifelines also signals new mobile spaces in which to explore debates about the place of cultural identity in the digital age. Key debates and discussions within policymaking for satellite radio in Canada moved from emphasizing the advantages of circulating content throughout the space of the footprint, and how this spatial advantage is superior to the limited range of a terrestrial broadcast tower, to a focus on how a single satellite radio service can establish a presence across a variety of mobile devices and private leisure spaces. This shift is reconfigured by the ubiquity of mobile listening devices. Both the satellite radio and digital music industries have become focused on maintaining a presence within lifelines.

While Sirius's MySXM is no longer an available user feature, other online features like on-demand programs continue to be available, and the company has introduced video features and exclusive podcasts that are all added components to the broadcast service. At the time of MySXM's introduction, it was stressed that the "company ha[d] always seen satellite radio as its core mission and the Internet as a value-added service," and then-CEO Mel Karmazin called Internet radio and its business model "'a race to the bottom,' noting that the company add[ed] features like MySXM because customers want[ed] them, 'not because we think it's good business sense.'"[96] Not long after Karmazin offered these comments, however, he stepped down as CEO, in December 2012, and was replaced by interim CEO James E. (Jim) Meyer. After Karmazin's exit, John Malone's Liberty Media would begin to exert more control over SiriusXM, and the company pursued further acquisitions in the internet audio market, namely Pandora internet radio. In October 2012 Karmazin said, "My instincts today are that Liberty does not need me." Liberty, at that time, controlled 49.5 percent of the satellite company, but expectations were that it would soon take control of SiriusXM.[97] The next and final chapter discusses Liberty Media's increasing control of SiriusXM and the company's ongoing expansion in programming and in the pursuit of subscribers, and what this means for the value of music in the streaming space age.

7 Embedded Radio and the Limits of Expansion

For over a century radio has facilitated connections between ears, minds, voices, and music. In *Selling Sounds*, David Suisman describes how songs were crafted "specifically, deliberately, and essentially as commercial products," although "it would be a mistake to suggest that the commercial production of songs rendered consumers passive and inert."[1] Similarly, Simon Frith succinctly explains that music is both a means of expression and a commodity;[2] the same can be said of radio. Within this spectrum, between culture and commodity, listeners find a sense of themselves. The voices and songs that cycle throughout the day, week, month, and year are the lifeblood of radio. Over SiriusXM's history, the satellite radio service has changed and expanded alongside new technologies and shifts in listening practices. It has done so in the range of channels available as well as in the range of situations and contexts within which people listen. In the ongoing search for connection and meaningful musical experiences, subscription satellite radio promises to be the source of these moments; whether or not it always delivers on this promise is another story.

Over SiriusXM's history, the cultural technology of the satellite has remained distinct as an object of aspiration and wonder. In SiriusXM's more recent history, the world of the satellite feels closer to human perception, due in part to the rise of what has been termed *space capitalism*.³ Outer space is a new frontier for control and colonization by those few with the means to enter orbit, like billionaires who made their fortunes through tech and entertainment companies. Space is the domain for those hoping to expand capital beyond Earth in the never-ending pursuit of growth.⁴ In July 2021 Virgin's Richard Branson was the first billionaire to launch himself into space with a commercial spaceflight company.⁵ The year before, Elon Musk's SpaceX launched astronauts into space, the first time astronauts had entered orbit on a privately owned spacecraft (prior instances were either suborbital or financed by the Russian government). The SpaceX launch was described by *Forbes* as "a culmination of a decades-long effort to transform space into a new frontier of entrepreneurship."⁶ SpaceX has also launched SiriusXM satellites into space. Musk and other space billionaires, like Amazon's Jeff Bezos, have competing "space visions." Though they differ, they share the alarming notion that space can be "designed for wealthy people like themselves."⁷ These examples symbolically reflect the decades-long fantasy that the economy can expand forever, thanks in part to new technological developments.⁸ The satellite, for instance, has expanded the more limited scope of music selections on Earth-born broadcast radio. In many ways, figures like Elon Musk, Richard Branson, and Jeff Bezos and their pursuit of space as future places to inhabit, tour for leisure, and extract resources from feel unique to the early 2020s. But as the story of satellite radio indicates, outer space has been the backdrop for the pursuit of audiences-as-subscribers and for crafting ideas about prestige radio and the perceived value it gestures toward. The growth of online and streaming music listening, coupled with the ongoing significance of outer space as a frontier for ideas about the expansion of capital, suggest that we are in a sort of streaming space age, in which technologies from the space age continue to drive idealistic visions of never-ending expansion in profits, subscribers, and resources in the streaming age.

SiriusXM has ordered new satellites from Maxar Technologies, with the twins SXM-11 and SXM-12 set to launch around 2026–2027. SXM-9 and SXM-10 were also ordered from Maxar, in 2021, and all four of these satellites are part of a third-generation constellation. They are meant to replace existing capacity and use more powerful antennas to cover Canada and southern Alaska.⁹ While internet radio and streaming media have become dominant forms of listening for music consumers, the figure of the satellite remains central to SiriusXM's business model. The same can be said for the automobile despite recent efforts to encourage out-of-vehicle listening. The company now competes with dashboard options like Apple CarPlay and Android Auto, which bring features like music streaming apps from the user's smartphone to the vehicle through display and voice input. Following the COVID-19 pandemic lockdowns, changes to the nature of work and the workforce mean more people now work from home and commute less.¹⁰ The space of the automobile remains invaluable but uncertain. SiriusXM is now available on devices like Amazon's Alexa and via custom apps on Google Chromecast, smart TVs, Roku, Sonos, and video game consoles like Sony PlayStation.¹¹ In 2019 SiriusXM debuted an "Essential" subscription plan with hopes to reach more audiences outside of the car. It costs $1 a month for the first three months and then $8 a month after this trial period. The plan includes all music and talk programs aside from certain sports channels and Howard Stern.¹² In the late summer of 2021 an ambitious multimedia campaign by SiriusXM promoted out-of-car listening, indicating the company's realization that the future of satellite radio will be determined by modes of listening beyond the space of the automobile. In fact, *RAIN News* reported that this was SiriusXM's "largest-ever" media campaign, which included a number of celebrities such as Kevin Hart, Bebe Rexha, and Dave Grohl.¹³ The campaign unfolded online as well on billboard spaces and advertisements in airports in New York and San Francisco. In one television and online advertisement, SiriusXM host and comedian Kevin Hart is informed by musicians and SiriusXM hosts LL Cool J, Bebe Rexha, and Kehlani that he does not need to listen to his own channel in the car but can use his phone, laptop, and Amazon Alexa smartscreen in the home while he walks on a treadmill and bakes in the kitchen. As Hart learns about each device that can be used to listen to his Laugh Out Loud Radio, he loudly exclaims, "What!?" The emphasis of the ad is that one can listen *anywhere*.¹⁴ That said, a big part of SiriusXM's future is a "hybrid" radio option in vehicles, one that involves both its internet service, including Pandora, and the satellite broadcast. The campaign to remain dominant on the dashboard remains.

This concluding chapter discusses the significant changes to SiriusXM after the exit of Mel Karmazin in February 2013. Jim Meyer replaced Karmazin until December 2020, when he stepped down and was replaced by Jennifer C. Witz, who was previously the director of sales. Karmazin's exit happened at a time when Liberty Media was gaining more control over the company and the online features of SiriusXM were becoming more pronounced. Throughout the 2010s and 2020s SiriusXM has increasingly competed with its streaming music competitors, and the satellite radio service has become focused on not only continuing to communicate perceived value through celebrity talent and a notion of prestige radio programming, music and otherwise, but also on ensuring it functions as a sort of embedded radio. Satellite is a type of radio that is central to everyday life in the streaming era and in a context of the platformization of music. It's available everywhere and with programming for an ever-growing list of taste cultures and audience types, who are targeted via new service offerings, new modes of datafication, and new mergers and acquisitions that aim to expand its programming and subscriber base.

LIBERTY LOOMS LARGE

SiriusXM is now one of the three divisions of the Liberty Media conglomerate's ownership stakes: SiriusXM, Formula One, and the Atlanta Braves. In October 2012 the monopoly satellite company had more than 22 million subscribers and was projected to add 1.8 million more by the end of year. An article in *Billboard* reflected on Liberty's initial move into SiriusXM, and said that Liberty Media's founder (and now chairman) John Malone is known for risky bets, but that the "deal to take a 40% stake in Sirius is, by his own admission, one of his best [moves]." The article continued, "At the peak of the financial crisis in 2009, a desperate Mel Karmazin, CEO of Sirius, needed a $530 million loan to help the company avoid bankruptcy. Malone lent Sirius the cash at an eye-watering 15% and attached warrants that eventually gave Liberty Media a 40% stake in the [company]." In October 2012 that 40 percent was worth $4.2 billion.[15] Along with SiriusXM, Malone owned more than one-quarter of Live Nation Entertainment, the biggest concert promoter and leading ticketing company (following the merger of Live Nation and Ticketmaster), and in 2012 Malone secured an option to increase Liberty's share in Live Nation to 35 percent.[16]

In 2013, in a corporate reorganization SiriusXM Holdings Inc. replaced SiriusXM as the publicly held corporation, under which SiriusXM became a wholly owned subsidiary.[17] In March, before Meyer was chosen as CEO, *Variety* reported that the company was in a strong financial position, with excess cash on hand. SiriusXM did not need to spend aggressively on growing the business because the merger and other talent acquisitions over the years had placed the company in a secure position. As such, it was planning a significant stock buyback process.[18] Thanks to strong automotive sales, this moment marked a post-merger record for net subscriber additions, adding 715,000 in the second quarter and reaching a total of more than 25 million.[19] The company boasted that it was the largest radio company in the world by revenue for the third consecutive year, with revenue of $3.8 billion at the end of 2013.[20]

In early 2014 *Billboard* reported that Liberty was working to fully absorb SiriusXM but was getting pushback due to a takeover price that many felt was too low ($3.68 per share or around $10.6 billion). Liberty owned 53 percent of the company at this time. Meyer defended this price by emphasizing that the company was now competing with web-connected cars.[21] SiriusXM's revenue continued to grow, and reports in February 2014 indicated that fourth-quarter revenue jumped 12 percent, from $892 million to $1 billion, with another 411,000 subscribers added. Increased expenses and the need to pay down debt, however, meant that overall yearly earnings were down.[22] In late March Malone withdrew his bid for a total takeover. But Liberty Media ownership of SiriusXM dramatically increased over the next few years. In 2015 Liberty ownership was 62.86 percent, and it was projected that by 2021 the company would own, directly and indirectly, over 80 percent of the outstanding shares of common stock.[23] As of 2023, Liberty owned 82.4 percent of SiriusXM.

Liberty ownership brought SiriusXM under the same corporate umbrella as live music ticket sales monopoly Live Nation Entertainment. SiriusXM, then, must be considered within a broader context of conglomeration and concentration across the music industries. Many of the world's music industries are increasingly affected by an oligopoly of transnational music, technology, and financial companies. With respect to record labels and music publishing, the "Big Three"—Universal Music Group, Sony Music Group, and Warner Music Group—dominate the record industry.[24] And while there is a long history of legitimate concern about media concentration, as well as myriad examples of uneven power dynamics between the record industry and recording artists and numerous instances of exploitation and precariousness in the cultural and creative industries,[25] these ongoing issues have been intensified through financialization, new streaming technology, and platformization. These new forces are tightening the grip of power held by the biggest companies across the media industries.[26] In the transition from music industries largely organized around material copies of recorded music to one in which digital formats are dominant, or as communication scholar Patrick Burkart describes, the transformation of the music industry into a service industry, major labels have retained their power and have been influential in these changes.[27] Even as companies like Spotify have grown in use and subscribers, the bulk of revenues are paid out to record labels.[28] While major labels and Big Tech companies have garnered significant power and influence in the music industries, concern about the sustainability of music careers and artists being fairly compensated for their work has grown more pronounced.[29] SiriusXM's pursuit of growth in its subscriber base and in becoming a key component in the Liberty Media conglomerate can be seen as aligned with a broader context of alarmingly high levels of concentration across the media industries.

SUBSCRIPTION MUSIC RADIO IN THE STREAMING SPACE AGE

Like broadcast radio, SiriusXM is embedded in the music industries. It is also a quintessential media company of the streaming era because of the subscription model it has used since its inception. As Andrew Bottomley argues, radio has been a trailblazer for the modern streaming media era,[30] and Sirius and XM are without question a key aspect of this trajectory. In 2013 Apple's dominance in digital music downloads was no longer such an advantageous position because subscription access enabled users to navigate extensive music catalogs through streaming services. Music subscriptions, at that time, were said to have "only scratched the surface,"[31] and as one analyst reported, "Apple's billion-dollar music question is how to embrace streaming while maintaining dominance in downloads."[32] Two years later, Apple would spark excitement and enthusiasm over the launch of Beats 1 (later rebranded as Apple Music 1), a radio station available via Apple Music, which came after the $3 billion acquisition of Beats Electronics in 2014. Beats 1 would borrow from the SiriusXM playbook by emphasizing access to music "trendsetters" like Zane Lowe from BBC Radio 1, but also musicians like Dr. Dre, St. Vincent, and Elton John.[33] Nearly fifteen years after the launch of Sirius and XM, Apple turned to a subscription model for music listening. In these ways, SiriusXM has been a key player in the platformization of music, even as it continues to rely on characteristics of broadcast radio to distinguish itself from major streaming platforms.

SiriusXM's subscription model, and its strategy of using celebrities and big name music personalities to draw in subscribers, enabled the company to be well-positioned to compete with Apple Music and Spotify despite the cheaper subscription plans of these new competitors. Apple Music also had the advantage of being embedded in the Big Tech industry with revenues from device sales. But SiriusXM had the advantage of being an established player in the record industry, continuing a long legacy of radio being a place where new music is introduced to consumers for the purpose of selling records, concert tickets, and merchandise. The trade press coverage of Sirius's and XM's early years often highlighted new celebrity hosts and related program offerings. In the mid-2010s, however, there was a noticeable shift to cover SiriusXM as a force for breaking, or promoting, new artists to subscribers. To be sure, radio has been an outlet for showcasing new music for a long time. The record industry has relied on radio to develop new artists and to reach a listening public. Those records selected to play on air have driven album sales and shaped talent acquisition policies at labels.[34] In some cases, as with formats like Top 40, radio has bolstered hit songs and sparked chart activity, and in others, as with college and campus radio, it has programmed music not heard elsewhere, often by local and independent artists.[35] Radio has had the ability to shape tastes and direct listeners to music in a way that has added context and insight, helping listeners navigate "an insistent and ubiquitous marketplace for music."[36] SiriusXM has amplified this reciprocal relationship.

Throughout the 2010s, a number of profiles of programmers, hosts, and channels made reference to hit songs or artists that generated extensive airplay on SiriusXM before breaking elsewhere. One profile of Scott Greenstein, president and chief content officer, said that the company "remains a key outlet to break artists," and it mentioned Sam Hunt and Brandy Clark as two artists who landed label deals in 2014 after getting attention from SiriusXM's country channels.[37] Vice president of music programming for pop formats Kid Kelly was featured a number of times in *Billboard* over this span of years. Kelly started with Sirius in 2003 and was influential in playing tracks that then generated high record sales. Kelly has said that "programming should be 'gut first.' 'We need more visionaries who make pop an exciting format with new artists all the time.'"[38] Speaking about pop group One Direction's 2011 song "What Makes You Beautiful," senior VP of promotion from Columbia Records Lee Leipsner said that Kelly's Hits 1 channel on SiriusXM first played the song before other radio stations did, a move that Kelly attributed to programmers taking more risks on satellite radio.[39] The pop trio AJR had a brief hit with "I'm Ready" in 2013, a song that Kelly had said demonstrated "real potential" and was "very different from what's available on the radio right now." SiriusXM airplay led to an uptick in iTunes sales and "an extremely strong" response by SiriusXM listeners, a fact made evident via SiriusXM's "very secret metrics."[40] In trade press coverage about some of the more niche music channels on SiriusXM, a similar trajectory unfolded over these years. On the electronic and dance channels, Jonathan "Geronimo" Broth, VP of music programming for electronic and dance formats, said in June 2016 that over the previous few years these channels had focused prominently on emerging artists.[41] Further, the buzz around indie rock band Parquet Courts in 2013 was attributed to a South by Southwest spot, touring in Europe, and airplay on SiriusXM.[42] On the country channels, Chase Rice's label "bidding war" was ignited in part due to SiriusXM. Rice had a song, "Ready, Set, Roll," that hit number twenty-seven on the Hot Country Songs chart without terrestrial airplay and helped Rice's EP sell forty-two thousand copies in less than two weeks "largely thanks to SiriusXM" airplay.[43] And on hard rock Octane, program director Vincent Usuriello said that the channel "moves the needle": "You instantly see a reaction in sales, streams or views when we start playing a song."[44] This emphatic fixation with being the *first* place to hear new artists or songs indicates a developing strategy to position SiriusXM against its competitors like Spotify and Apple Music in the realm of introducing new music and demonstrating value to the record industry. Subscribers may also feel a sense of perceived value in having access to songs or artists first.

In the streaming age, competition between SiriusXM and music platforms has been fierce. At the end of 2022, Spotify became the first music streaming service to reach two hundred million paid subscribers, a feat that was announced around the same time that the company stated it would cut 6 percent of its workforce (approximately six hundred employees). Apple's latest figures at that time came from its 2019 annual report, which indicated a total of sixty million paid subscribers.[45] As Robert Prey, Marc Esteve Del Valle, and Leslie Zwerwer explain, "Spotify is often called the 'new radio' for the influence it has on breaking new songs and artists."[46] But SiriusXM is not willing to easily concede its *new* radio status. By remaining a vital outlet for the record industry and by providing subscribers with access to a wide range of music and artists and advanced access to "the next big thing," SiriusXM aims to weather the competition from streaming music services. Whether or not it can remain competitive against these Big Tech juggernauts is a question that only time will answer. But at the moment of writing, SiriusXM has expanded through strategic acquisitions, namely Pandora; the building of new studios; and the signing of new on-air talent.

WHEN TWO BECOME ONE, AGAIN: SIRIUSXM ACQUIRES PANDORA INTERNET RADIO

By 2015 numerous music format radio stations were simulcasting their airplay online, and streaming music had surpassed digital downloads for the first time as the record industry's largest revenue source.[47] Commercial FM radio was still facing criticism for its limited playlists and predictable programming. Radio companies like iHeartMedia and Cumulus Radio were cutting staff "in the face of crippling debt."[48] Radio listenership had fallen but was still relatively stable in 2016, at fourteen hours per week in the US, down from twenty hours in 2007. Tommy Boy Records founder

Tom Silverman said, "People in their 50s aren't going to stop listening to radio—but teens who haven't listened to radio never will."[49] As the popularity of internet radio grew, it was no surprise that SiriusXM would become a major investor in internet radio competitor Pandora. Investing in an already existing internet radio service allowed SiriusXM to move into the internet radio market in substantial ways and to streamline overlapping aspects of the two radio companies. In June 2017 SiriusXM invested $480 million in Pandora, taking one-third of its board seats.[50] Following the investment, executive VP/general counsel of SiriusXM Patrick Donnelly said that this move could lead to major growth opportunities because Pandora was "much more in the ad-supported radio business.... Plus, they've got 75 million 19- to 36-year-olds, which is a great market we don't touch."[51] In an interview with Pandora's president and CEO Roger Lynch (who was new in the role in 2017 after founder Tim Westergren resigned), a few challenges facing Pandora were detailed: the internet radio company was not faring well against on-demand streaming services like Spotify and Apple Music, and although its free tier had 76 million users, only 390,000 were premium users, and only 4.9 million were subscribers. The year before Pandora had faced stagnating listener growth, and this inspired the launch of an enhanced ad-free tier called "Pandora Plus," which it was hoped would encourage more fan engagement.[52] Because of ongoing financial losses, 19 percent of the company was sold to SiriusXM. Lynch stressed that Pandora had a big share of the audio ad marketplace and that his approach would be to look for synergies with SiriusXM that would benefit both companies.[53]

Some analysts agreed that the two companies coming closer together had clear benefits for both parties, and in September 2018 a deal to merge the two companies was on the table. *Billboard* asked, "How would the combination affect record labels, publishers and music creators?" The move to acquire Pandora in an all-stock transaction was said to be predictable after SiriusXM had taken a 19 percent stake in the company. Some benefits of the merger were said to be a growth in overall users and subscribers and the abilities for each service to promote the other, to have Pandora's personalization features become part of the SiriusXM "experience," to fund upgrades and technology improvements, and to enhance SiriusXM's position in the radio market.[54] Programming could be shared between both services, and SiriusXM's established position in the automobile would increase Pandora's in-car distribution. One label executive, however, expressed concern with Liberty's moves and asked whether the eventual plan would be to roll SiriusXM, Pandora, and Live Nation "into one gigantic music conglomerate with divisions in many music sectors."[55] Nevertheless, on February 1, 2019, the $3.5 billion purchase of Pandora was finalized. SiriusXM also took ownership of the ad technology platform AdsWizz, a Pandora subsidiary.[56] The acquisition was said to have created "a digital-radio behemoth with customized radio playlists, a contract with Howard Stern and an estimated 100 million listeners who tune in from cars, smartphones and laptops alike."[57] Meyer said that the goal was to have free Pandora users become subscription listeners for SiriusXM. And if a subscriber were to leave SiriusXM, they would ideally find themselves listening to the ad-supported Pandora.[58] Meyer added that the company should "never lose a listener" and be a "strong ally" to artists growing a fan base or to a label breaking a new artist.[59] One graphic shared by *RAIN News* depicts upsells moving from Pandora to SiriusXM and churn moving in the opposite direction.[60] The deal also meant that SiriusXM would have increased control over Pandora, as a number of high-ranking Pandora employees were now out of the picture, including CEO Roger Lynch and CFO Naveen Chopra.[61]

After the acquisition of Pandora, a major artist partnership was announced at the end of 2019. Drake, one of the biggest streaming artists at that time, partnered with SiriusXM. The two companies had a combined reach of around one hundred million listeners, and Drake was convinced of the benefits to his own career and profile. Greenstein said that this partnership had "more content and marketing components than the others that came before: 'It is unequivocally the deepest we've ever gone on any artist deal.'" Other massive announcements that followed the merger included U2 X-RADIO, a deal with LeBron James's media company, and a multiyear deal with Marvel Comics to create original podcasts. Programming from all these deals would be available on both SiriusXM and Pandora.[62] A "Pandora-powered" channel was in the works, one that could be based on a user's favorite artist, and Pandora would have more "curated branded stations," borrowing from the SiriusXM playbook.[63] By acquiring Pandora, SiriusXM gained a large number of users and potential

Figure 19. SiriusXM satellite beaming down on Planet Pandora. *Source:* Steve Knopper, "Will Labels Have a Sirius Problem?," *Billboard*, February 9, 2019, 22.

subscribers. This enhanced the reach of the combined services, attracted new talent, and enabled the company to remain a powerful player in the music streaming and internet radio market.

Following the Pandora acquisition, SiriusXM symbolically announced plans to open a new studio in Los Angeles, in Hollywood to be precise. This new studio would reflect the company's presence and growth in key music and entertainment centers, complementing its studios in New York, Washington, and Nashville. The new LA studio (seven times the size of its old one in LA) was said to include a large performance space, and its opening was celebrated with three LA-based live broadcasts of Howard Stern's talk show ("a radio event if there ever was one," according to *Variety*).[64] Greenstein said that LA had "such an overly concentrated group of musicians, artists, actors, authors, comedians . . . and now they can use this facility instead of having to fly to New York."[65] Some hinted that the LA base could be a great spot to reach a new generation of talent after Stern retires. One example of LA-based growth at SiriusXM following the opening of the new studio was the $150 million acquisition of Team Coco, talk host and comedian Conan O'Brien's podcast company.[66] O'Brien entered into a five-year talent agreement with SiriusXM. As part of the deal, SiriusXM created a new channel, Team Coco Radio.[67] After the debut of the channel, a number of guest-hosted shows aired across music channels, including one on the 90s alternative channel Lithium, that involved O'Brien playing bands performing early in their careers on his talk show.

SiriusXM continued to invest in, or acquire, companies. In 2020 it acquired the podcasting company Stitcher for $325 million, invested $75 million in SoundCloud, and acquired podcast distribution and analytics company Simplecast. The company also set its sights on acquiring radio behemoth iHeartMedia, a rumor that sparked a group of consumer rights advocates and artists' rights organizations to petition the US Department of Justice. They said such a move would have a "likely catastrophic" effect on radio markets.[68] Liberty was given approval to increase its ownership stake in iHeartMedia, but the company has since sold all its stock in the radio company. Podcasting companies Stitcher and Simplecast became part of SiriusXM's growing presence in the podcast market, an area where advertising revenue could be increased and the company's established track record of acquiring big name audio talent could be put to further use. Marvel inked a multiyear podcast deal; Fox News personality Megyn Kelly moved her podcast to SiriusXM; and some SiriusXM talk radio shows became podcasts that were available on Pandora, including a number by Fox News and ones hosted by Ricky Gervais, Kevin Hart, and Jason Ellis. After the Stitcher deal, *RAIN News* reported that SiriusXM now claimed the largest addressable audience in the US across all digital audio, with 150 million listeners.[69] These moves placed SiriusXM in direct competition with other major media companies like Spotify, which had signed a large deal with Joe Rogan for his podcast, and *The New York Times*, which had bought Serial Productions.[70] In April 2022, podcasting listening was said to have grown 266 percent in share of listeners over five years, namely in the eighteen to thirty-four age demographic.[71] These acquisitions and developments will enable SiriusXM to reach a larger base of listeners who are younger and who may not be willing to pay for the full in-car subscription for SiriusXM, at least not initially.

The Pandora acquisition can be characterized as moving SiriusXM outward from its primary demographic of high-income, automobile-owning subscribers to cast its *potential-subscriber* net much wider. The company has said that through marketing efforts and the inclusion of radios in lower-priced vehicles, SiriusXM was reacting to the changing demographics of its subscriber base, including an increase in customers from the millennial generation.[72] Certainly acquiring the more than seventy million Pandora users helps the company to accomplish this goal. New channels also reflect this turn to younger audiences, such as two new rock channels that premiered in 2017 and play music from the 1990s and 2000s, called Turbo and PopRocks.[73] These years are nostalgic for millennial listeners and, as chapter 3 explains, tying programming to nostalgia has been a key aspect of the channel lineup and the company's programming strategy. Some of the genre-specific channels have also broadened their playlists because a younger generation is less accustomed to the packaging of music by genre, as it might be organized in a record store, and instead is more familiar with a wide range of music being presented on a screen. Jeff Regan, the senior director of music programming for Alt Nation, has brought in rappers like Lil Peep and Post Malone to the "still largely rock-based playlist."[74] While not reflective of every SiriusXM channel, these moves raise questions about how the early goal in the satellite radio industry, of having distinct music channels for niche styles of music, will fare in this new media environment, where the aspiration of continued growth may be reducing niche focus in favor of broader appeal, or at least retooling the balance between these two styles of music programming. In the effort to attract younger subscribers, programming strategies shift from music discovery taking place vertically and deeply, within a genre-specific channel, to a broader, horizontal formation, in which exploration occurs across expanding channel options and platforms, as well as a wider variety of music selections within a given channel.

SATELLITE RADIO AND THE RECORD INDUSTRY IN THE TWENTY-FIRST CENTURY

The radio industry and the record industry remain closely connected, even as both industries undergo notable shifts, changes, challenges, and new business models. The point at which SiriusXM acquired Pandora is a good one for taking stock of the era of Liberty ownership and SiriusXM's role within the larger music and record industries. The company exerts its influence through its ability to strike content deals with major artists, its royalty rate negotiation power, its ability to break new songs and artists, its connecting with listeners and subscribers across artist- and genre-based channels, and as a platform with international reach. In 2016 Greenstein said, "Labels have seen that if we get behind a record, we can force terrestrial [radio] to follow."[75] SiriusXM's relevance to the record industry is perhaps not more evident than in the fact that Taylor Swift performed in the SiriusXM New York studio on the date of the release of her album *Reputation* in 2017, the best-selling album of that year. More recently, a pop-up Taylor Swift channel (Taylor's Version) coincided with the release of *The Tortured Poets Department* (2024). SiriusXM's *value* to the record industry is evident in "pop-up" channels that intensely feature the music of one artist. The Kenny Chesney channel No Shoes Radio began as a three-month feature on XM Radio. Chesney, an American country music singer-songwriter, was involved in organizing the channel, suggesting features like a phone line for listeners to leave messages. The channel was successful in terms of listenership and new subscriptions, and it eventually became an online radio channel before moving to channel 59.[76] In another example, a Bon Jovi Radio channel was launched alongside a larger *What About Now* album and tour campaign in the summer of 2012.[77] Further, the ongoing expansion of channels allows SiriusXM to continually focus in on certain genres and artists in ways that deliver "the right songs in front of the right users," a practice that Darrin Smith, vice president of music programming, calls "hypertargeting."[78] This connects to chapter 2's discussion of the connection between a satellite and its targeting capabilities in the context of the military and surveillance and then that of targeting subscribers. In more recent years, through datafication and the growth of online and digital services by satellite radio, there has been a marked shift from targeting to hypertargeting, which has taken place alongside the expansion of music programming.

The practice of integrating platforms like YouTube and SoundCloud into the channel lineup also generates synergies between SiriusXM and major tech platforms. What platforms do, according to media scholar Robert Prey, is circulate and curate existing content, as opposed to producing or owning it, inaugurating "relations of dependency among creators and the industries they draw upon."[79] While SiriusXM and Pandora do not strictly fit most definitions of platforms, namely because they do not currently enable users to upload clips or content to their apps or web pages, the move to enhance and grow online offerings must be understood as a reaction to, or acknowledgment of, the increasing prominence and power of platforms across the media and entertainment industries. The platformization of culture extends into the realm of music circulation. Nick Srnicek explains that at a general level, "platforms are digital infrastructures that enable two or more groups to interact.... More often than not, these platforms also come with a series of tools that enable their users to build their own products, services, and marketplaces."[80] As chapter 6 explains, the platformization of cultural production involves the increasing influence of digital platforms on the cultural industries and on the organization of labor.[81] SiriusXM has partnered with major platforms such as YouTube and TikTok to create channels that bring aspects of these platforms, and ideally their users and potential subscribers, into the purview of satellite radio. In 2015 senior vice president and general manager of music programming and digital music Steve Blatter said that a "new partnership with YouTube to co-produce shows and share data is key to talent discovery. 'We're able to see what's bubbling under on YouTube at an incredibly early stage.'"[82] One example of this is a weekly spotlight on the top deephouse tracks on YouTube playlists, called Chill Trending Tracks-Powered by YouTube, which was a partnership initiated by Ben Harvey, program director for the Chill, Pop2K, and BPM channels.[83] After the Pandora acquisition, this coming together of platform and programming only intensified. And with the rollout of one hundred new Xtra channels in 2019, SiriusXM subscribers could personalize stations using algorithm-generated content, thanks to the merging of SiriusXM's technology with Pandora.[84]

 SiriusXM's relationship to the record industry and broader music industries indicates ongoing synergies between these companies and industries, as well as SiriusXM's efforts to balance a notion of freedom and experimentation with the drive to acquire subscribers in its music programming. A focus on experimentation was especially apparent in the early days of XM and Sirius. But the pursuit of profits and ongoing competition with similar music services have, in some cases, lessened the more experimental and niche aspects of music programming by satellite radio. The relevance of SiriusXM to the wider music industries has resulted from decades of music programming that highlights eras, decades, and more niche styles of music that have great potential to connect with listeners and fans in intimate and meaningful ways. At the same time, it has become a company that the record industry and other music businesses depend on, increasingly due to its acquisition of other media and music companies. The ongoing relevance of radio to the record industry is evident across SiriusXM's programming and operations. As such, it makes sense that the satellite radio service continues to use record industry genre categories, as well as decades and artist brands, to structure its music programming. SiriusXM is evidence of the continued relevance of radio and legacy media and the power that major entertainment conglomerates retain in the streaming media era.

"TURNING YOUR ORBIT AROUND"

The pursuit of endless growth is unstable and impossible, despite attempts by the heads of Big Tech and entertainment companies to explore, and extract from, outer space. Countless media companies have collapsed under the weight of working to appease shareholders year after year. And SiriusXM is no exception to being disciplined by these pressures. In December 2022 the company announced plans to fire staff due to "faltering sales growth." Analysts made sense of the move to reduce staff amid a broader shift in the tech world, where concerns about a recession led to companies like Amazon and Facebook cutting more than ten thousand employees each. At the end of 2022, SiriusXM had forty million subscribers, roughly the same number that it had three years earlier.[85]

The Pandora acquisition, in terms of subscriber numbers, was also not a move that led to unchecked growth for the Liberty Media conglomerate. In February 2020 the satellite radio company had added 355,000 subscribers in the fourth quarter but lost 88,000 Pandora listeners.[86] By the end of 2020, reports were that Pandora had continued to lose subscribers and that it had incurred a $976 million impairment charge due to royalty payment costs. SiriusXM executives were still publicly optimistic about the potential for the Pandora acquisition to grow the company in podcasts and in strengthening its position in the automobile.[87] This sort of optimism is strategic when it comes to assuring shareholders of future growth potential.

In the years following the merger of Sirius and XM, a slight shift in emphasis from music to talk programming, for the sake of enhancing advertising revenue, was evident in annual reports and trade reporting. After the acquisition of Pandora, the ad-based revenue stream was more aggressively pursued by SiriusXM. Pandora has allowed advertisers to "target and connect with listeners based on various criteria including age, gender, geographic location and content preferences."[88] Hypertargeting listeners enables the company to increase its revenues through advertising. As subscriber growth slowed over 2022, some analysts reported that the satellite radio company had turned its attention to podcasts and live sports to increase its ad revenue potential, with advertising from sports accounting for nearly a quarter of ad sales. CEO Jennifer C. Witz said that this would help the company to "lure and retain more premium subscribers."[89] As with the acquisition of Pandora and the sophisticated collection of metrics and listener data that came with it, SiriusXM's forays into the podcasting industry helped to grow its share of advertising revenue and overcome the limits of subscriber growth. In 2021 SiriusXM announced the merger of three internal sales divisions (Sirius, Pandora, and Stitcher) to create a unit called SXM Media.[90] The new ad-sales unit, with a combined reach of 150 million listeners, would go on to sign Crooked Media, a podcast publisher known for *Pod Save America*;[91] unveil an interactive podcast browser for advertisers called SXM Podcast Universe (described as "a spacey interactive display of podcasts in the SXM network, and their audience characteristics");[92] and create branded podcast episodes, in which sponsors purchase an "episode takeover" that has brand affiliation without outright and more obvious marketing.[93] In January 2022 SXM Media introduced AudioID, "a new way of identifying, describing, and targeting listeners and groups of listeners." According to Chris Record, the senior vice president of technology and operations at SXM Media, AudioID reflects a "new era of identity," one in which we are not defined "by who we are on paper or the cookies we leave behind, but by our interests and passions."[94] The notion of hypertargeting is central to SiriusXM in the 2020s, as it looks beyond subscriber numbers to continually grow the company across listeners, revenues, and programming. But subscriber revenue continues to be the most significant revenue source for SiriusXM. In 2020 subscriber revenue for SiriusXM was just over $5.8 billion, and advertising revenue was $157 million. Pandora's 2020 numbers included subscriber revenue of $515 million and advertising revenue of nearly $1.2 billion.[95] In 2021, across the broader audio entertainment environment, subscription revenue was nearly two-thirds of overall audio revenues.[96] Even with subscriber revenue remaining the biggest piece of the pie, it's clear that the purchase of Pandora increased the opportunity to earn more through advertising while the satellite radio music programming remained free of advertising.

In the vehicle, the future seems to lie in SiriusXM's 360L radios. These are described as a "hybrid satellite + streaming solution [that] transforms the listening experience for customers by introducing more personalized content and recommendations, and, in the future, will enable [SiriusXM] to improve [its] targeted advertising."[97] This platform maintains a satellite link to vehicles while also enabling two-way connectivity through wireless networks.[98] The goal was to have these radios in 80 percent of new SiriusXM-equipped vehicles by 2025, but the company has said this mark will be missed by a few years. The interface was initially introduced in 2018 and installed in Dodge Ram trucks. In addition to combining the satellite and streaming services, the 360L interface will provide the company with data to enable it to understand subscriber activity and to more effectively market services to consumers. According to Witz, the 360L platform will boost interactivity but also help to overcome the fact that the "broadcast" service's "pipe is only so big." 360L allows for "an unlimited amount of content."[99]

Driven by changes in technology, listening habits, and new industry competitors, SiriusXM and its expansion in programming and online spaces reflect the *embeddedness* of media as well as the embeddedness of music in our daily lives (the description of AudioID reinforces this point). Although music is still organized and programmed in studios and sent out via satellite or internet connection, we maintain close connections with music through mobile devices from the car to a smartphone. These lifelines are maintained and sustained through digital infrastructure, which, as sociologist Susan Leigh Star describes, has the property of embeddedness: "Infrastructure is sunk into and inside of other structures, social arrangements, and technologies. People do not necessarily distinguish the several coordinated aspects of infrastructure."[100] Sociologist and communications scholar Phillip N. Howard theorizes an embedded media perspective, one that indicates communication technologies have become deeply embedded in our personal lives. This allows for a high degree of user control (the sort of thing SiriusXM has tried to accomplish to some extent with personalized and customizable channels), and it exemplifies the fact that a small "elite owns, manages, produces, and channels information through media such as television, radio, and newspapers."[101] An embedded media perspective, according to Howard, can be measured according to fit, status, and link. "Status" situates people as both producers and consumers of information, and "link" connects different spheres of life more efficiently and effectively than traditional media. SiriusXM, and its hybrid status of having elements of traditional or legacy media as well as new media characteristics of internet radio and customizable stations, fits these two categories in some ways but also fails to do so in others. However, on the topic of "fit," satellite radio demonstrates embeddedness. Fit pertains to that which suits our daily routines.[102]

Music, and music programmed by radio, is a part of routine, and it shapes and structures the cycles of our lives. Media scholar Paddy Scannell writes about how a sense of *dailiness* indicates a level of meaningfulness in the activities of radio and television. Broadcasting, according to Scannell, articulates our sense of time; that cyclical time, against the linear time of life from birth to death, is about the routines and the repetitions of life: the "repetitive cycles of days, weeks, months and years." Scannell asks, "Do we escape the arrow of time, our being-towards-death in these ways?"[103] Radio routines and habits, adds scholar Anne Leonora Blaakilde, "are embodied practices, established in connections and conjunctions with material-cultural actors in flux and they are transformable over time."[104] Even in SiriusXM's relatively short history, different modes of providing a sense of dailiness and routine through music have been introduced and established. These modes have been shaped by technologies like the vehicle and smartphone, and the ways we interact with and listen to music have been altered in conjunction with the ways we structure our daily lives.

In music's close connection to identity, it too is a form of embeddedness. Music comes in and out of our lives in waves and cycles. Media industries, like satellite radio, determine how and when and in what context this takes place. Our tastes, preferences, and knowledge of certain types of music form an orbit around us, and media technologies make the connections between songs. Through radio, as this book has argued, music is given a sense of value. SiriusXM and Spotify share the fact that they are two of the few major companies in the business of music circulation in the streaming era, but they differ with respect to how they work to craft the perceived value of their music. The former relies much more heavily on talent acquisition, on radio voices, and on the names and familiarity of genres and their affiliated artists. The latter relies on the algorithm, some element of human curation, and the playlist as indicators of value.[105] SiriusXM also feels more invested in the record industry model of organizing music, by genre and album and artist, versus moods and experiences, though this has shifted over time as well, particularly in the ever-expanding channels available via the SiriusXM app. Both now rely on subscribers, though SiriusXM is a precursor in this regard.

When the SXM-7 satellite was being tested in orbit, it experienced certain failures that instigated an evaluation of the technology. In these failures and disruptions, the hidden infrastructure and the behind-the-scenes components became apparent. SiriusXM is faced with the task of not appearing broken to its subscribers. In the quest for new promises to make and deliver to shareholders, it must retain a commitment to the sort of music programming that listeners value, or its future may be compromised. Over the course of the histories of Sirius, XM, and SiriusXM, a commitment to music programming that offers a listening experience *beyond* that of commercial broadcast radio has been a key reason listeners subscribe. But the task of finding new subscribers each year, across categories of age, class, taste, and so forth, presents its own challenges and can compromise the experimental and novel aspects of programming. Through an array of music channels, the company has aimed to reach both niche and mass tastes, but how far can innovation go when the goal is to continually raise subscriber numbers and profits? Is one media company with so much influence in the audio sector able to retain values of variety and diversity in its program offerings? This very question was raised when Stitcher announced it was being discontinued as of August 2023 because SiriusXM was incorporating podcasts into its subscription business. Responding to the announcement, one Twitter user wondered if they would still be able to find the more obscure podcasts in the Sirius universe.[106] With Liberty Media becoming a major player in radio, as well as in ticketing and venues through its stake in Live Nation, what does this mean for the variety of music that circulates across the industries and outlets in the contemporary media environment? Outer space is a realm sought after with aspirational ideas of unlimited growth, but the reality of the situation is much more down to earth. Ongoing issues facing the satellite radio industry include paying fair royalties to artists, crafting music programming that is both varied and unpredictable while also of benefit to the hit-making needs of the record industry, competing for space in vehicles and on smartphones, and having the money to acquire talent. These issues all have the ability to shape and alter the perceived value of music programming by satellite radio.

To conclude, it's worth reinforcing the resilience of radio in digital spaces. SiriusXM shows how radio remains influential in the record industry as well as in communicating meaning to listeners through songs. In April 2023 Russ Crupnick, the managing partner of MusicWatch, said that Sirius was well positioned to continue growing its subscriber base and compete with other major music streaming companies. The top reasons for retaining a subscription to SiriusXM were said to be the ability to use it all the time and everywhere, its good value, and its role in facilitating music discovery. It's unique because of its "legacy advantages" in the dashboard and its "SXM Everywhere" capabilities.[107] Though the company's business model has changed as it has looked to talk programming and podcasts as key areas of growth in the early 2020s, it still invests heavily in music. It remains connected to sounds and genres from the past but also to the cycles of record production and the circulation of new music. It's a service that has adapted with the introduction of new technological realities in order to remain embedded in listeners' daily lives. A feature like the SiriusXMU Faculty is an example of this: XMU listeners log on to complete a survey that ranks frequently played songs in order to get a sense of which ones should stay around and which ones should go. Taking the survey in February 2023, I was given a total of thirty songs to rate.

This feature reflects the increasing prominence and influence of the platformization of music but also retains SiriusXM's legacy media structure of doing the programming for audiences, as opposed to having users upload content. The one hundred commercial-free Xtra channels that SiriusXM introduced in 2019 also include functions that mirror music platforms, like the skip-ahead feature. Finally, a Channels You Might Like feature aims to "drive a more personalized listening experience in the car for new users via SiriusXM with 360L."[108] Sirius and XM launched at a time when the companies could position themselves against the shortcomings of commercial broadcast radio. And satellite radio's continued relevance extends from the fact that a subscription model for a "premium" radio service is enticing compared to the sort of "freemium" tiers of listening that companies like Spotify have been known for (and from which they are increasingly trying to get away). With a love for music comes the justification for paying for subscription access, as long as the programming is aligned with expectations of value, something that changes over time.

This chapter's final section is titled "Turning Your Orbit Around." It's a line from the lyrics to Wilco's "Jesus, Etc." (recall that this song was released around the time of Sirius's and XM's debuts). It has been called "the gentlest, most accessible song on . . . one of the best indie rock albums of all time."[109] In a few of the song's choruses, Jeff Tweedy sings, "Tuned to chords strung down your cheeks / Bitter melodies turning your orbit around." Music can force us to turn our own orbits around. To reorient ourselves. The song goes on to claim that "our love is all we have," and by extension perhaps, so too are the songs that communicate these emotions, moving in cycles across and throughout our limited time in this world. We can become one with song, if we follow this logic. In this way, music is embedded in our lives and in our bodies, and this reinforces the appeal of media technologies, like satellite radio, that effectively bring us the music we desire. The history of satellite radio sheds light on the whens and hows of our connections with music, how we find the songs we love and what institutions are behind the processes by which music becomes embedded in our lives and memories.

Others have connected this song to the September 11 attacks in New York City. The song was recorded before the events of September 11, but it was released one week afterward, on September 18, 2011, as a free download on Wilco's website (it would later be released on Nonesuch Records on April 23, 2002). Writing about the song, *Rolling Stone* said, "Calling down the redemptive power of love and music with verses that anticipated the imagery of 9/11, Jeff Tweedy's finest moment was the right medicine at precisely the right time."[110] The disconnect/connection between the song and event exemplifies the imperfect place of music and memory, but also music's role as a means to recollect and as an emotional layer to historical moments. And as this book has revealed, this is a historical moment defined by tense relationships between the national and the global, the expanding world of the internet and online culture, changes to music technologies and listening practices, and the centralization and standardization of broadcasting. It is an era in which the satellite, through the field of music radio, takes on the cultural significance of enhanced choice in music and national, then international, coverage. A radio beyond radio. A type of listening that feels closer to the songs and artists we love, despite the satellite's distance. Something more meaningful, but not without its limits and shortcomings. As is music, as is life.

Notes

INTRODUCTION

1. *Billboard* Hot 100 formed when radio-play and singles sales charts merged in 1958.
2. Many of Elvis's hits were released before the Hot 100 was formed.
3. Breihan, *Number Ones*, 4.
4. Lee Abrams, "Willie Nelson Travelogue and XM Thinking from the Early Days," *Lee Abrams Media Visions*, April 9, 2007, www.leeabramsmediavisions.com/blog/willie-nelson-travelogue-and-xm-thinking-from-the-early-days.
5. Frank Saxe, "WolrdSpace Aims to Bring Satellite Radio to the Planet," *Billboard*, August 19, 2000, 91.
6. "Heavenly Music," *The Economist*, March 16, 2002, www.economist.com/technology-quarterly/2002/03/16/heavenly-music.
7. Carla Hay and Angela King, "September 11 and the Economy Force Radio to Rethink Itself," *Billboard*, December 29, 2001, 65.
8. Electronic Frontier Foundation, "RIAA v. The People: Five Years Later," *EFF*, September 30, 2008, hwww.eff.org/wp/riaa-v-people-five-years-later; and Amy Forliti, "Single Mom Can't Pay $1.5M Song-Sharing fine," *NBC News*, November 5, 2010.
9. Phyllis Stark, "Napster Is Talk of Town Meeting," *Billboard*, November 25, 2000, 39.
10. An advertisement in *Billboard*, November 25, 2000, 4.

11. Carla Hay and Angela King, "September 11 and the Economy Force Radio to Rethink Itself," *Billboard*, December 29, 2001, 65.

12. Hay and King, "September 11 and Economy," 65.

13. "Introducing XM Radio Commercial (2002)," posted May 3, 2017, by Vhs Vcr, YouTube, www.youtube.com/watch?v=UJVitlGEfl4.

14. "Sirius Satellite Radio Commercial (2005), posted January 2, 2021, by TV-inNEWFOUNDLAND2, YouTube, www.youtube.com/watch?v=TGMahshSamo.

15. "XM[:] The Early Days," posted June 9, 2017, by Art Vuolo, YouTube, www.youtube.com/watch?v=9fAeLdYQOTw.

16. XM Satellite Radio, *2000 Annual Report*, 2001, 1.

17. XM Satellite Radio, *2000 Annual Report*, 1.

18. XM Satellite Radio, *2001 Annual Report*, 2002, 1, 7.

19. Marc Schiffman, "Pie in the Sky: After Near Crashes, Satellite Radio Set to Soar," *Billboard*, June 7, 2003, 1.

20. Rushkoff, *Get Back in the Box*, 2-3.

21. Lee Abrams, "NPR, Paul's Deal, and Young Listeners—How to Attract Them . . . Or Is It Even Possible? . . . and Guitars." *Lee Abrams Media Divisions*, March 26, 2007, www.leeabramsmediavisions.com/blog/npr-pauls-deal-and-young-listeners-how-to-attract-themor-is-it-even-possible-and-guitars.

22. Lee Abrams, interview by Brian Fauteux, June 13, 2023.

23. Weisbard, *Top 40 Democracy*.

24. Berman, *This Book Is Broken*, 55; and Dahlman, "'Big Beautiful Mess,'" 74.

25. Dahlman, "'Big Beautiful Mess,'" 61; and Klein, "'New Radio,'" 474.

26. For more on freeform radio and college radio see Kramer, *Republic of Rock*, and Jewell, *Live from the Underground*.

27. Mark Richardson, "Albums: In the Aeroplane Over the Sea," *Pitchfork*, September 26, 2005, https://pitchfork.com/reviews/albums/5758-in-the-aeroplane-over-the-sea.

28. Carl Wilson, "Are We Finally Ready to See Neutral Milk Hotel for What It Really Was?," *Slate*, March 1, 2023, https://slate.com/culture/2023/03/neutral-milk-hotel-aeroplane-sea-meaning-jeff-mangum.html.

29. April Ludgate, "Parks and Rec's April Ludgate Celebrates the Anniversary of Neutral Milk Hotel's Finest Work," *Vulture*, February 11, 2013, www.vulture.com/2013/02/april-ludgate-on-neutral-milk-hotels-in-the-aeroplane.html.

30. Katie Atkinson, "Aaliyah's 20 Best Songs: Staff List," *Billboard*, August 25, 2021, www.billboard.com/media/lists/best-aaliyah-songs-9619318.

31. Hunt, *Measuring Time, Making History*, 14.

32. Reynolds, *Retromania*, x-xi.

33. Academic debates about postmodernism and pastiche had considered this much earlier and are discussed further in chapter 3.

34. Phyllis Stark, "Napster Is Talk of Town Meeting," *Billboard*, November 25, 2000, 41.

35. Gavin Edwards, "Satellite Radio Prepares for Liftoff," *Rolling Stone*, September 13, 2001, 101.

36. Prindle, "No Competition," 306.

37. Prindle, "No Competition," 308.

38. Peter DiCola and Kristin Thomson, "Radio Deregulation: Has It Served Citizens and Musicians? A Report on the Effects of Radio Ownership Consolidation Following the 1996 Telecommunications Act," *The Future of Music Coalition*, November 18, 2002, 3–4.

39. "Heavenly Music," *The Economist*, March 16, 2002, www.economist.com/technology-quarterly/2002/03/16/heavenly-music.

40. Amanda Mull, "The 2000s Never Ended," *The Atlantic*, December 27, 2019, www.theatlantic.com/health/archive/2019/12/america-still-living-2000s/604174.

41. Olivia O'Bryon, "These Early 2000s Trends Are Back for 2022—For Better or Worse," *Forbes*, December 23, 2021, www.forbes.com/sites/oliviaobryon/2021/12/23/if-2002-is-any-indication-these-style-trends-will-be-hot-in-2022/?sh=30c39d582b4d; and Hannah Ewens, "Trends Used to Come Back Round Every 20 Years: Not Anymore," *Vice*, December 14, 2022, www.vice.com/en/article/bvmkm8/how-the-20-year-trend-cycle-collapsed.

42. Neda Ulaby, "From Tumblrcore to 2014core, the Nostalgia Loop Is Getting Smaller and Faster," *NPR*, March 1, 2022, www.npr.org/2022/03/01/1081115609/from-tumblrcore-to-2014core-the-nostalgia-loop-is-getting-smaller-and-faster.

43. See, for instance, Jameson, "Postmodernism"; and Hutcheon, *Politics of Postmodernism*.

44. Tim Ingham, "Over 73% of the US Music Market Is Now Claimed by Catalog Records, Rather Than New Releases," *Music Business Worldwide*, January 6, 2022, www.musicbusinessworldwide.com/over-82-of-the-us-music-market-is-now-claimed-by-catalog-records-rather-than-new-releases2.

45. Scannell, *Radio, Television and Modern Life*.

46. Lee Abrams, interview by Brian Fauteux, June 13, 2023.

47. Innis, *Empire and Communications*; and Jenkins, Ford, and Green, *Spreadable Media*, 37.

48. Tsing, "Global Situation," 331, 336.

49. Straw, "Circulatory Turn," 23.

50. Kornbluh, *Immediacy*.

51. Taylor, "Circulation, Value, Exchange," 258.

52. Taylor, "Circulation, Value, Exchange," 265

53. Brunet, "Introduction," 13.

54. Taylor, *Making Value*, 14. Taylor draws on anthropological literature on value and highlights two historical trajectories: Marx writing on forms of capitalist value and Marcel Mauss on gift exchange. Taylor also provides an overview of noncapitalist and nongift conceptions of value.

55. Adorno, *Culture Industry*; and Adorno, "On Popular Music," 303.
56. Clover, *Roadrunner*, 27–28.
57. Clover, *Roadrunner*, 32.
58. Ed Christman, "Scott Greenstein," *Billboard*, July 26, 2014, 20, 22.
59. Morris, "Music Platforms and Culture," 2.
60. Holt and Perren, "Introduction," 1.
61. Holt and Perren, "Introduction," 5.
62. Williams, *Television*.
63. Editorial Collective, "Welcome to Media Industries," 1.
64. Acland, *American Blockbuster*, 274.
65. Thorburn and Jenkins, introduction, 12.
66. Gitelman, *Always Already New*.
67. Acland, "Introduction"; Hamilton, "Unearthing Broadcasting"; and Edgerton, *Shock of the Old*.
68. XM Satellite Radio Holdings Inc., *Form 10-K (Annual Report): Filed 3/15/2001 for Period Ending 12/31/2000*, 2001, 1.
69. XM Satellite Radio Holdings Inc., *1999 Annual Report*, 2000, 4.
70. Sirius XM Holdings Inc., *United States Securities and Exchange Commission, Washington, D.C. 20549: Schedule 14A, Proxy Statement Pursuant to Section 14(a) of the Securities Exchange Act of 1934*, 2020.
71. Eriksson et al., *Spotify Teardown*, 87.
72. "Muzak Uses Satellite to Send Music," *Billboard*, October 4, 1980, 111.
73. Russo and Kirkpatrick, "Beyond the Terrestrial?," 156, 168.
74. Robyn Wells, "'Off Button' Is Key Competition," *Billboard*, March 13, 1982, 34.
75. Jean Callahan and Doug Hall, "Commission's Report Triggers Battle over NPR Network Future," *Billboard*, February 10, 1979, 20.
76. Jean Callahan, "Psychological Changes Go with Satellite," *Billboard*, April 19, 1980, 33.
77. Jan Dawson, "Bucking the Content Trend," *Variety*, November 22, 2016, 26.
78. Glenn Peoples, "Does Listening Time Matter?," *Billboard*, September 14, 2019, 16.
79. Gary Trust and Silvio Pietroluongo, "The Satellite Prophet," *Billboard*, January 25, 2014, 22–27.
80. Hilmes, "Foreword," iii.
81. In *Tunes for All?*, Michelsen et al. offer an overview of research on music radio, highlighting influential fields, authors, books and articles, and concepts. One noted influence within this trajectory is the growth of popular music studies into a distinct field of study in the 1990s, which involved a number of researchers whose main interest was pop music radio. Michelsen et al. refer to Anglo-American perspectives that shaped the study of music radio, such as social history and cultural studies, highlighting Paddy Scannell and David

Cardiff's 1991 book, *A Social History of British Broadcasting*, vol. 1, *1922–1939*, which included two chapters devoted to music and radio. For other studies of music radio, both new and old, see Michelsen et al., *Music Radio*; Baade, *Victory through Harmony*; Coddington, *Hip Hop Became Hit Pop*; Klaess, *Breaks in the Air*; Kramer, *Republic of Rock*; Jewell, *Live from the Underground*; Russo, "Radio Formats;" Johnson, "Date with the Duke"; Fauteux, *Music in Range*; and Berland, *North of Empire*.

82. Dan Papscun, "Swifties Start Wave of Ticketmaster Monopoly Scrutiny," *BNN Bloomberg*, November 18, 2022, www.bnnbloomberg.ca/swifties-start-wave-of-ticketmaster-monopoly-scrutiny-1.1848601; David McCabe and Ben Sisario, "Justice Dept. Is Said to Investigate Ticketmaster's Parent Company," *New York Times*, November 18, 2022, www.nytimes.com/2022/11/18/technology/live-nation-ticketmaster-investigation-taylor-swift.html; and Nilay Patel, "Taylor Swift vs. Ronald Reagan: The Ticketmaster Story," *The Verge*, March 21, 2023, www.theverge.com/23645057/taylor-swift-ticketmaster-eras-tour-beyonce-antitrust-monopoly-reagan-senate-hearing-congress.

83. Drott, *Streaming Music, Streaming Capital*, 198.

84. For more on the financialization of media, see deWaard, *Derivative Media*.

85. Acland, *American Blockbuster*, 29.

86. Acland, *American Blockbuster*, 30

87. Richard Hogan, "Layoffs Hit Hollywood Reporter, Billboard Magazine," *AdWeek*, April 16, 2015, www.adweek.com/performance-marketing/hollywood-reporter-billboard-layoffs; and Matt Donnelly, "Layoffs Hit Hollywood Reporter-Billboard Media Group," *Variety*, January 10, 2019, https://variety.com/2019/music/news/hollywood-reporter-billboard-layoffs-1203105070.

88. Eriksson et al., *Spotify Teardown*, 7.

1. THE SONGS DOWN TO EARTH

1. Ozzi, *Sellout*.
2. Clarke, preface, vi–vii.
3. Pierce, *Beginnings of Satellite Communication*, 4–5.
4. Pierce, *Beginnings of Satellite Communication*, 5.
5. Rees, *Satellite Communications*, 83.
6. Parks and Schwoch, introduction, 10.
7. Rees, *Satellite Communications*, 26.
8. "List of Satellites in Geostationary Orbit," *Satellite Signals*, www.satsig.net/sslist.htm.
9. "Boeing 702 Satellites Solar Arrays Possibly Defective," *Space and Tech*, September 28, 2001, https://web.archive.org/web/20071226225020/http://www.spaceandtech.com/digest/flash2001/flash2001-082.shtml.

10. Edmund L. Andrews, "F.C.C. Plan for Radio by Satellite," *New York Times*, October 8 1992, D1, D17.

11. "Robert Briskman Appointed Chairman and CEO," *Satellite News*, June 1, 1992; and Bethany McLean, "Satellite Killed the Radio Star," *Fortune*, January 22, 2001, https://money.cnn.com/magazines/fortune/fortune_archive/2001/01/22/295563/index.htm.

12. Cynthia Littleton, "Radio Satcasters Tuning in Some High-Profile Investors," *Variety*, November 30, 1998, 21.

13. "Sirius Gets $150 million Loan," *R&R*, March 16, 2001, 4.

14. Billboard Staff, "Marley Song Launches XM Satellite Radio Feed," *Billboard*, September 26, 2001, www.billboard.com/music/music-news/marley-song-launches-xm-satellite-radio-feed-78299.

15. XM Satellite Radio, *2001 Annual Report*, 2002.

16. Hilmes, "Foreword," v.

17. Jeffrey Yorke, "CD Radio Sues XM on Three Counts of Patent Infringement," *R&R*, February 26, 1999, 1, 16.

18. "Heavenly Music," *The Economist*, March 16, 2002, www.economist.com/technology-quarterly/2002/03/16/heavenly-music.

19. Jeffrey Yorke, "XM Raises Add'l $235 Million, Stock Yo-Yos," *R&R*, July 14, 2000, 4.

20. "Staff or No Author. Pay-Radio Firm Attracts $250 Million in Investments," *Los Angeles Times*, June 9, 1999, www.latimes.com/archives/la-xpm-1999-jun-09-fi-45601-story.html.

21. "Business Briefs: Sirius Satellite Launch Delayed," *R&R*, January 1, 2000, 4.

22. Marilyn A. Gillen and Frank Saxe, "Idealive Gets Funding for 1st Indie Artist," *Billboard*, July 15, 2000, 74.

23. Sirius Satellite Radio, *Annual Report 2002, Form 10-K: Annual Report Pursuant to Section 13 or 15(d) of the Securities Exchange Act of 1934, for Fiscal Year Ended December 31, 2002*, 2003, F-16.

24. Jeffrey Yorke, "Sirius: Three to Get Ready," *R&R*, December 8, 2000, 1, 19.

25. Sirius XM Radio Inc., *Form 10-K (Annual Report): Filed 02/25/10 for the Period Ending 12/31/09*, 2010, 4.

26. Sirius XM Radio Inc., *Form 10-K (Annual Report): Filed 03/10/09 for the Period Ending 12/31/08*, 2009, 7.

27. Trevor Paglen, "Some Sketches on Vertical Geographies," *e-flux Architecture* (October 2016), www.e-flux.com/architecture/superhumanity/68726/some-sketches-on-vertical-geographies.

28. Gunter D. Krebs, "XM 1, 2 (XM Rock, Roll)," *Gunter's Space Page*, https://space.skyrocket.de/doc_sdat/xm-1.htm.

29. XM Satellite Radio Holdings Inc., *XM Satellite Radio Inc. Form 10-K: Annual Report Pursuant to Section 13 or 15(d) of the Securities Exchange Act of 1934, for the Fiscal Year Ended December 31, 2006*, 2007, 7.

30. XM Satellite Radio, *Form 10-K: Annual Report for December 31, 2006*, F-39.

31. XM Satellite Radio Inc., *Form 10-K (Annual Report): Filed 4/2/2003 for Period Ending 12/31/2002*, 2003, 21; and Sirius XM Holdings Inc., *Form 10-K (Annual Report): Filed 02/04/14 for the Period Ending 12/31/13*, 2014, 11.

32. Sirius XM Holdings Inc., *Form 10-K: Annual Report Pursuant to Section 13 or 15(d) of the Securities Exchange Act of 1934, for the Fiscal Year Ended December 31, 2020*, 2021.

33. Sirius XM Holdings Inc., *Form 10-K: Annual Report Pursuant to Section 13 or 15(d) of the Securities Exchange Act of 1934, for the Fiscal Year Ended December 31, 2016*, 2017, 4.

34. Sirius XM Holdings, *Form 10-K: Annual Report Pursuant to Section 13 or 15(d) of the Securities Exchange Act of 1934, for the Fiscal Year Ended December 31, 2020*, 6.

35. Glen Dickson, "A Sirius New Facility," *Broadcasting & Cable*, November 29, 1999, 40, 42.

36. Bethany McLean, "Satellite Killed the Radio Star," *Fortune*, January 22, 2001, https://money.cnn.com/magazines/fortune/fortune_archive/2001/01/22/295563/index.htm.

37. Glen Dickson, "A Sirius New Facility," *Broadcasting & Cable*, November 29, 1999, 40, 42.

38. Bottomley, *Sound Streams*, 36.

39. Frank Saxe, "XM Opens HQ, Plans Production Of Receiver," *Billboard*, September 30, 2000, 86.

40. XM Satellite Radio, *2000 Annual Report*, 2001, 13.

41. "The XM Satellite Radio System," *Broadcast Engineering*, December 2002, 64.

42. XM Satellite Radio, *2000 Annual Report*.

43. "Heavenly Music," *The Economist*, March 16, 2002, www.economist.com/technology-quarterly/2002/03/16/heavenly-music.

44. XM Satellite Radio Inc., *Form 10-K (Annual Report): Filed 4/2/2003 for Period Ending 12/31/2002*, 2003, 27.

45. Steve Traiman, "DVD Audio, Satellite Advances Could Transform Car Listening," *Billboard*, May 22, 1999, 87.

46. Glen Dickson, "A Sirius New Facility," *Broadcasting & Cable*, November 29, 1999, 40, 42.

47. Steve Traiman, "DVD Audio, Satellite Advances Could Transform Car Listening," *Billboard*, May 22, 1999, 87.

48. Chuck Taylor, "The Radio of the Future," *Billboard*, November 28, 1998, 67.

49. XM Satellite Radio Inc., *Form 10-K: Annual Report Pursuant to Section 13 or 15(d) of the Securities Exchange Act of 1934, for the Fiscal Year Ended December 31, 2005*, 2006, 10.

50. Marc Schiffman, "Pie in the Sky: After Near Crashes, Satellite Radio Set to Soar," *Billboard*, June 7, 2003, 82.

51. Sirius Satellite Radio, *Form 10-K: Annual Report Pursuant to Section 13 or 15(d) of the Securities Exchange Act of 1934, for Fiscal Year Ended December 31, 2002*, 2003, 8.

52. Sirius initially planned for 105 repeaters in forty-six cities and XM for 1,700 in seventy cities. See Don Waller, "Long Road to Market," *Variety*, August 14, 2000, 30.

53. "FCC's Silver Anniversary: No Fanfare, Just a Look at the Record," *Broadcasting*, June 15, 1959, 66.

54. "Will Satellites Jam Spectrum?," *Broadcasting*, October 14, 1957, 62.

55. Rees, *Satellite Communications*, 26.

56. Elliot P. Fagerberg, "A Review of the ITU Conference," *Broadcast Engineering*, January 1963, 22.

57. Collis, "Geostationary Orbit," 52.

58. "XM Satellite Technology," *Rohde & Schwartz*, www.rohde-schwarz.com/ca/technologies/satellite-broadcast/xm-satellite/xm-satellite-technology/xm-satellite-technology_55613.html; XM Satellite Radio Holdings Inc., *Form 10-K: Annual Report Pursuant to Section 13 or 15(d) of the Securities Exchange Act of 1934, for the Fiscal Year Ended December 31, 2008*, 2009, 3; and Sirius Satellite Radio Inc., *Form 10-K (Annual Report): Filed 02/29/08 for the Period Ending 12/31/07*, 2008.

59. "Method and Apparatus for Audio Output Combining: US Patent Issued on July 11, 2006," *Patent Storm*, https://web.archive.org/web/20070926220151/http://www.patentstorm.us/patents/7075946-description.html.

60. "FCC Approves Nationwide Satellite-Radio Plan," *The Gazette* (Cedar Rapids-Iowa City), March 4, 1997, P1.

61. Thomas Hazlett, "Local Motives: Why the FCC Should Scrap Its Absurd Rules for Satellite Radio," *Slate*, March 16, 2004, https://slate.com/culture/2004/03/the-fcc-s-absurd-rules-for-satellite-radio.html.

62. Sirius Satellite Radio Inc., *Form 10-K (Annual Report): Filed 3/12/2004 for Period Ending 12/31/2003*, 2004, 12.

63. Sirius Satellite Radio, *Form 10-K (Annual Report) for 12/31/2003*.

64. Michael Wolff, "Cruise Control," *New York Magazine*, September 25, 2000, https://nymag.com/nymetro/news/media/columns/medialife/3823.

65. Berland, *North of Empire*, 12

66. Mike Boyle, "Industry News," *CMJ New Music Report*, June 24, 2002, 6.

67. Mike Boyle, "Industry News," *CMJ New Music Report*, July 8, 2002, 6.

68. XM Satellite Radio Holdings Inc., *Form 10-K (Annual Report): Filed 3/16/2000 for Period Ending 12/31/1999*, 2003, 11.

69. "Satellite OK; Opera On TV," *Billboard*, November 25, 1978, 42.

70. Mildred Hall, "Jazz on NPR: Satellite Transmission Coming as Network Looks into Future," *Billboard*, May 28, 1977, 65.

71. Parks and Starosielski, introduction, 1, 4.
72. Shields, "Bridge Spanning Past," 346.
73. Parks and Starosielski, introduction, 7.
74. Paglen, "Some Sketches on Vertical Geographies."
75. Peters, *Speaking into the Air*, 206.
76. Peters, *Marvelous Clouds*, 166.
77. Arendt, *Human Condition*, 1.
78. Parks, "Satellites, Oil, and Footprints," 125.
79. "Music in Space: We Are the World," *The Economist*, January 30, 2021, 70.
80. Peters, *Speaking into the Air*, 246–247, 215.
81. Tony Taylor, "Music for a Stellar Generation," *Los Angeles Times*, November 25, 1989, B6.
82. XM Satellite Radio, *2003 Annual Report*, 2004.
83. Taylor, "Music and Rise of Radio," 425–426.
84. XM Satellite Radio, *2001 Annual Report*, 2002.
85. XM Satellite Radio Holdings Inc., *1999 Annual Report*, 2000, 1.
86. Bethany McLean, "Satellite Killed the Radio Star," *Fortune*, January 22, 2001, https://money.cnn.com/magazines/fortune/fortune_archive/2001/01/22/295563/index.htm.
87. Radio News Bureau, "'Electronics' Is Still the Magic Word," *Television Digest with Electronics Reports*, August 13, 1955, 1.
88. Matthew J. Culligan, "Radio—The Evolving Medium," *Television Magazine*, January 1959, 45.
89. "Anniversary Greeting," *Broadcasting*, January 11, 1960, 76.
90. "Stereo-Slanted 1958 NAMM Show Passed 10,000-Visitor Mark," *The Billboard*, August 11, 1958, 11, 44.
91. Jane Blanksteen, "Heavenly Music: Mercury Whistles, Uranus Clicks, Pluto Drums," *The Globe and Mail*, May 7, 1979, 21.
92. David Cheal, "Lunar Tunes: How Music Responded to the Era of Space Travel," *FT.com*, July 17, 2019, www.ft.com/content/57e2b32c-a49c-11e9-974c-ad1c6ab5efd1.
93. "'Beeps' on Records," *The Billboard*, October 24, 1960, 10.
94. Taylor, *Strange Sounds*, 72.
95. Taylor, *Strange Sounds*, 73–74.
96. "*Music for Heavenly Bodies*," *Billboard*, July 28, 1958, 22.
97. "*Music from Outer Space*," *Billboard*, August 11, 1962, 49.
98. "Satellites & Sacks Inspire Cleffer Muse," *Billboard*, May 19, 1958, 1.
99. "Sales Aids from Philco," *The Billboard*, June 24, 1957, 28.
100. Taylor, *Strange Sounds*, 76.
101. "Belgian Ops Flying Right with Telstar," *Billboard*, August 18, 1962, 49.
102. "Research on 'Whistlers': Stanford Studies Weird Music of Outer Space," *Los Angeles Times*, December 7, 1954, 6.
103. "Voices of the Satellites!," *Billboard*, November 10, 1958, 24.

104. "A Child's Introduction to Outer Space," *Billboard*, October 19, 1959, 38.

105. McLeod, "Space Oddities," 337.

106. Bob Rolontz, "Jazz A Popular Idiom Today—Why?," *The Billboard*, June 29, 1959, 30.

107. Youngquist, *Pure Solar World*, 141, 135.

108. Corbett, *Extended Play*, 7.

109. Veal, "Starship Africa," 455–456.

110. Jon Savage, "Odysseys and Oddities: Jon Savage Compiles the Definitive Space-Rock Tape," *MOJO*, March 1995, www.rocksbackpages.com.login.ezproxy.library.ualberta.ca/Library/Article/odysseys-and-oddities-jon-savage-compiles-the-definitive-space-rock-tape.

111. McLeod, "Space Oddities," 346.

112. McLeod, "Space Oddities," 346.

113. Mark Binelli, "Lil Wayne: Rap's Alien Genius," *Rolling Stone*, April 16, 2009, www.rollingstone.com/music/news/lil-wayne-raps-alien-genius-20090416.

114. McLeod, "Space Oddities," 346.

115. "Project Moon," *Billboard*, December 30, 1957, 35.

116. Taylor, *Strange Sounds*, 80–81; and Keightley, "Turn It Down!"

117. Susan Nunziata, "Satellite CD Radio Service Seeks FCC OK," *Billboard*, June 2, 1990, 88.

118. Douglas, *Inventing American Broadcasting*, 190–191.

119. Ashlee Vance, "A Ham Radio Weekend for Talking to the Moon," *New York Times*, June 26, 2009, www.nytimes.com/2009/06/27/technology/27moon.html.

120. "*Music for Heavenly Bodies*," *Billboard*, July 28, 1958, 22.

121. Marc Schiffman, "Tuned In: Radio," *Billboard*, June 5, 2004, 47.

122. Marc Schiffman, "Tuned In: Radio; Will FCC Crackdown Boost Sat-Casters?," *Billboard*, March 13, 2004, 81.

123. Dontay Thompson, "Geronimo Gets Sirius," *R&R*, October 31, 2003, 27.

124. Alex Ben Block, "TV without Pictures Gets a Hearing on Satellite Radio," *Emmy*, November/December 2006, 26.

125. Nick Biro, "Wholesalers Must Join Hands—Merc's Steinberg," *Billboard*, September 8, 1962, 5.

126. Bernard Chevry, "The 2nd International Music Industry Conference: Looking Ahead," *Billboard*, December 27, 1969, 158.

127. Claude Hall, "Aussie, U.S. RKO DJs Trade Mikes," *Billboard*, May 29, 1976, 1, 28.

128. David Farrell, "Regular FM/TV Simulcast Slated for Toronto Area," *Billboard*, November 4, 1978, 20, 102.

129. Mike Harrison, "The Complex Truth about Satellites," *Billboard*, September 5, 1981, 23.

130. Ray Herbeck Jr., "Satellite Radio Networking Looms—Distantly," *Billboard*, April 22, 1978, 32.

131. Advertisement, *R&R*, November 26, 1999, 5.

132. Parks, *Cultures in Orbit*, 2.
133. XM Satellite Radio Inc., *Form 10-K: Annual Report Pursuant to Section 13 or 15(d) of the Securities Exchange Act of 1934, for the Fiscal Year Ended December 31, 2005*, 2006, 13.
134. Parks and Starosielski, introduction, 5, 6.
135. Harvey, *Spaces of Global Capitalism*, 29.
136. Collis, "Geostationary Orbit."
137. Jarett, *Satellite Communications*, xiii.
138. Probst quoted in in Jaffe, *Satellite Communications*, 108.
139. Parks, *Cultures in Orbit*, 22, 29, 37.
140. "Visions Mobile Advertisement," *Music Week*, March 16, 1985, 43.
141. Stu Lambert, "SMS Dishes Up a Solution to Signal Sending Problem," *Music Week*, August 5, 1989, 29.
142. Peter Ludwig, "IDB Sets Digital Link for London, N.Y. Remotes," *Billboard*, April 9, 1988, 10, 14.
143. Auslander, *Liveness*, 77.
144. Peters, *Speaking into the Air*, 218.
145. Graham, *Vertical*, 25.
146. "'Project Moon' on Wax Orbit," *Billboard*, November 11, 1957, 24.
147. Paglen, "Some Sketches on Vertical Geographies."
148. Michael Wolff, "Cruise Control," *New York Magazine*, September 25, 2000, https://nymag.com/nymetro/news/media/columns/medialife/3823.
149. Berland, *North of Empire*, 134.
150. Berland, *North of Empire*, 136.

2. TARGETING SUBSCRIBERS IN SATELLITE RADIO'S FORMATIVE YEARS

1. XM Satellite Radio Inc., *Form 10-K: Annual Report Pursuant to Section 13 or 15(d) of the Securities Exchange Act of 1934, for the Fiscal Year Ended December 31, 2004*, 2005, 35.
2. Sirius XM Satellite Radio, *Proxy Statement and 2011 Annual Report*, 2012.
3. Matthew Benz, "Skepticism over Music Subscriptions: Net Services Plagued by Uncertainty," *Billboard*, June 22, 2002, 1, 80.
4. Styvén, "Intangibility of Music," 56.
5. Featherstone, "Automobilities," 10.
6. Eriksson et al., *Spotify Teardown*, 34.
7. "Set Parts Replacement Hits $900 Million, Says Baker," *Broadcasting Magazine*, December 30, 1957, 61.
8. Frank Saxe, "Big Audio Dynamite," *Billboard*, October 14, 2000, 101.
9. Marc Sciffman and Chuck Taylor, "Radio Eyes Potential of New Media," *Billboard*, September 18, 1999, 12.

10. Sciffman and Taylor, "Radio Eyes Potential of New Media," 116.
11. Chuck Taylor, "Format Worries, Cold Remedies Flavor Packed Radio Seminar/Awards Show," *Billboard*, October 23, 1999, 89.
12. Antony Bruno, "Got HD? Radio Format Starts New Push," *Billboard*, June 17, 2006, 16.
13. Sean Ross and Phyllis Stark, "Convention Capsules," *Billboard*, March 27, 1999, 90.
14. Marc Schiffman, "Tuned in: Radio," *Billboard*, March 15, 2003, 62.
15. CRTC, *FM Radio in Canada: A Policy to Ensure a Varied and Comprehensive Radio Service* (Ottawa: CRTC, 1975).
16. Hendy, "Pop Music Radio," 749.
17. Cwynar, "Brick, Mortar, and Screen."
18. See Razlogova, "Freeform Radio and Music Streaming"; and Jewell, *Live from the Underground*.
19. Williams, *Television*, 16.
20. Hilmes, "Radio and the Imagined Community," 351. See also Anderson, *Imagined Communities*.
21. Fauteux, *Music in Range*.
22. Parks, *Cultures in Orbit*, 78.
23. Parks, *Cultures in Orbit*.
24. Andrejevic, *Infoglut*, 1
25. Parks and Schwoch, introduction, 13.
26. Chow, *Age of the World Target*, 31.
27. Chow, *Age of the World Targe*, 35.
28. Kaplan, "Mobility and War."
29. Arditi, "Digital Subscriptions."
30. Manovich, *Language of New Media*, 55.
31. Andrejevic, *Infoglut*, 2.
32. Montell, *Cultish*.
33. Anna Washenko, "MusicWatch: Almost 11 Million Americans Illicitly Share Their Subscription Streaming Accounts," *RAIN News*, February 11, 2020, https://rainnews.com/musicwatch-almost-11-million-americans-illicitly-share-their-subscription-streaming-accounts.
34. Matthew Benz, "Skepticism over Music Subscriptions: Net Services Plagued by Uncertainty," *Billboard*, June 22, 2002, 1.
35. Ken Schlager, "How Many More Monthly Fees Can Consumers Stand?," *Billboard*, May 28, 2005, 6.
36. Matthew Benz, "Skepticism over Music Subscriptions: Net Services Plagued by Uncertainty," *Billboard*, June 22, 2002, 1.
37. Attali, *Noise*.
38. Israel Diamond, "Data Processing: The Music Scene," *Billboard*, January 17, 1970, M-5.

39. Kathryn Harris, "Master-Quality Transmission to Homes: Music Company Will Use Satellite to Bypass Record Store," *Los Angeles Times*, October 13, 1981, E1.

40. George Kopp, "Home Music Taping Via Cable Services," *Billboard*, May 2, 1981, 3, 80.

41. "Mail Order Radio on Air to Europe," *Music Week*, November 7, 1992, 5.

42. "Sony and Warner Back Digital Radio," *Music Week*, October 23, 1993, 3.

43. Annie Zaleski, "Four Columbia House Insiders Explain the Shady Math behind '8 CDs for a Penny,'" *AV Club*, June 10, 2015, www.avclub.com/four-columbia-house-insiders-explain-the-shady-math-beh-1798280580.

44. Formerly RCA Victor, which had operated a record club.

45. Eriq Gardner, "Columbia House, Former 'Music Club,' Files Chapter 11 Bankruptcy," *Billboard*, August 11, 2015, www.billboard.com/music/music-news/columbia-house-former-music-club-files-chapter-11-bankruptcy-6662989.

46. Matthew Stuart and Clancy Morgan, "How Columbia House Sold 12 CDs for as Little as a Penny," *Business Insider*, January 2, 2019, www.businessinsider.com/columbia-house-bmg-music-profit-money-business-model-2018-7.

47. Federal Trade Commission, Civil Action No. 2:23-cv-0932, Complaint for Permanent Injunction, Civil Penalties, Monetary Relief, and Other Equitable Relief, U.S. Dist. Ct., W. Dist. Wash. (June 21, 2023).

48. Annie Zaleski, "Four Columbia House Insiders Explain the Shady Math behind '8 CDs for a Penny,'" *AV Club*, June 10, 2015, www.avclub.com/four-columbia-house-insiders-explain-the-shady-math-beh-1798280580.

49. Ethan Trex, "It's a Steal! How Columbia House Made Money Giving Away Music," *Mental Floss*, June 21, 2011, www.mentalfloss.com/article/28036/its-steal-how-columbia-house-made-money-giving-away-music.

50. Matthew Stuart and Clancy Morgan, "How Columbia House Sold 12 CDs for as Little as a Penny," *Business Insider*, January 2, 2019, www.businessinsider.com/columbia-house-bmg-music-profit-money-business-model-2018-7.

51. Ethan Trex, "It's a Steal! How Columbia House Made Money Giving Away Music," *Mental Floss*, June 21, 2011, www.mentalfloss.com/article/28036/its-steal-how-columbia-house-made-money-giving-away-music.

52. Jack Hamilton, "Columbia House Offered Eight CDs for a Penny, but Its Life Lessons Were Priceless," *Slate*, August 12, 2015, https://slate.com/culture/2015/08/columbia-house-bankrupt-mail-order-cd-clubs-owner-finally-going-out-of-business.html.

53. Larry LeBlanc, "Canadian Retail Feels Record Clubs' Impact," *Billboard*, March 9, 1996, 52.

54. Ed Christman, "Record Clubs Come under Royalty Scrutiny," *Billboard*, April 6, 2002, 8B.

55. Ed Christman and Michael Poaletta, "Record Clubs Utilize New Strategies," *Billboard*, January 30, 1999, 1, 105.

56. Don Jeffrey, "Columbia House Readies for the Future," *Billboard*, February 19, 2000, 1, 84.

57. Brian Garrity and Don Jeffrey, "What Now for Col. House, CDnow?," *Billboard*, March 25, 2000, 1, 85.

58. Ed Christman, "Industry Outsider Makes Bid for Columbia House," *Billboard*, March 16, 2002, 16.

59. Brian Garrity. "Clubs Seek Better Access to Hits," *Billboard*, October 4, 2003, 5.

60. Michael Nelson, "Columbia House Is Finally Going Out of Business," *Stereogum*, August 11, 2015, www.stereogum.com/1823177/columbia-house-is-finally-going-out-of-business/news; and Brian Garrity, "BMG Direct Buys Rival," *Billboard*, May 21, 2005, 10–11.

61. Ed Christman, "Clubs Scale Back Freebies, Employees," *Billboard*, April 14, 2001, 3.

62. Jack Hamilton, "Columbia House Offered Eight CDs for a Penny, but Its Life Lessons Were Priceless," *Slate*, August 12, 2015, https://slate.com/culture/2015/08/columbia-house-bankrupt-mail-order-cd-clubs-owner-finally-going-out-of-business.html.

63. Eriq Gardner, "Columbia House, Former 'Music Club,' Files Chapter 11 Bankruptcy," *Billboard*, August 11, 2015, www.billboard.com/music/music-news/columbia-house-former-music-club-files-chapter-11-bankruptcy-6662989.

64. Ed Christman, "The Long Goodbye," *Billboard*, August 16, 2008, 12.

65. Bryan Bishop, "Columbia House, the Spotify of the '80s, Is Dead," *The Verge*, August 10, 2015, www.theverge.com/2015/8/10/9127703/columbia-House-mail-order-music-streaming-nostalgia.

66. Schiffman, "Tuned in: Radio," March 15, 2003, 62.

67. Mullen, *Rise of Cable Programming*, 94.

68. Streeter, "Blue Skies and Strange Bedfellows."

69. Marc Schiffman, "Tuned In: Radio: Will FCC Crackdown Boost Sat-Casters?," *Billboard*, March 13, 2004, 81.

70. Mullen, *Rise of Cable Programming*, 1.

71. XM Satellite Radio Inc., *Form 10-K: Annual Report Pursuant to Section 13 or 15(d) of the Securities Exchange Act of 1934, for the Fiscal Year Ended December 31, 2004*, 2005, F-19.

72. XM Satellite Radio Holdings Inc., *Form 10-K (Annual Report): Filed 3/16/2000 for Period Ending 12/31/1999*, 2000, 3.

73. Acland, *American Blockbuster*, 9.

74. "Oobu Joobu—Ecology," https://jpgr.co.uk/oobu5.html.

75. "Oobu Joobu Pt. 3," posted July 30, 2020, by Paul McCartney, YouTube, www.youtube.com/watch?v=AjL6emnJ1Rc.

76. Discussed in chapter 5.

77. Walt Love, "Sheridan's Black Satellite Network," *R&R*, February 4, 1983, 34.

78. XM Satellite Radio Inc., *Form 10-K (Annual Report): Filed 4/2/2003 for Period Ending 12/31/2002*, 2003, 9.

79. Smith-Shomade, "Narrowcasting in New World Information Order," 71.

80. Smith-Shomade, "Narrowcasting in New World Information Order," 78

81. Mullen, *Rise of Cable Programming*, 8.

82. Kristen Warner, "Blue Skies Again: Streamers and the Impossible Promise of Diversity," *Los Angeles Review of Books*, June 24, 2021, https://lareviewofbooks.org/article/blue-skies-again-streamers-and-the-impossible-promise-of-diversity.

83. Russo and Kirkpatrick, "Beyond the Terrestrial?," 169.

84. Marc Schiffman, "Tuned In: Radio," *Billboard*, November 8, 2003, 49.

85. XM Satellite Radio Holdings Inc., *XM Satellite Radio Inc. Form 10-K: Annual Report Pursuant to Section 13 or 15(d) of the Securities Exchange Act of 1934, for the Fiscal Year Ended December 31, 2007*, 2008, 38. In 2005, it was reported that Sirius counted unsold new cars equipped with its receivers *as* subscribers, reinforcing the centrality of the automobile to its business model. See Steven Strick, "Filtering New Music," *R&R*, November 4, 2005, 60.

86. Schiffman, "Tuned In: Radio," November 8, 2003, 49.

87. Jeffrey Yorke, "Radio Business: Sirius Listens to the Future," *R&R*, September 22, 2000, 4, 10.

88. Jeffrey Yorke, "Shares Slide as Sirius Struggles with Carmakers," *R&R*, April 6, 2001, 35; and Jeffrey Yorke, "Radio Group Heads Reveal Their Plans to Investors," *R&R*, June 8, 2001, 28.

89. Styvén, "Intangibility of Music."

90. Marc Schiffman, "Tuned In: Radio: U.S. Enjoys More Car Time," *Billboard*, October 25, 2003, 63.

91. Antony Bruno, "Sky's the Limit for Satcasters in 2005," *Billboard*, January 22, 2005, 39.

92. Marc Schiffman, "Tuned In: Radio: Ford Puts Sirius in the Driver's Seat," *Billboard*, January 15, 2005, 47.

93. XM Satellite Radio Inc., *Form 10-K (Annual Report): Filed 4/2/2003 for Period Ending 12/31/2002*, 2003, 45.

94. XM Satellite Radio, *2004 Annual Report*, 2005.

95. XM Satellite Radio Inc., *Form 10-K: Annual Report Pursuant to Section 13 or 15(d) of the Securities Exchange Act of 1934, for the Fiscal Year Ended December 31, 2005*, 2006, 37.

96. XM Satellite Radio Inc., *Form 10-K: Annual Report Pursuant to Section 13 or 15(d) of the Securities Exchange Act of 1934, for the Fiscal Year Ended December 31, 2004*, 2005, 18.

97. Sirius Satellite Radio, *2005 Annual Report and Proxy Statement*, 2006, 3.

98. Sirius XM Satellite Radio, *Proxy Statement and 2010 Annual Report*, 2011, 11.

99. Sirius Satellite Radio, *2006 Annual Report and Proxy Statement*, 2007, 2.

100. Tamara Conniff, "Stern's Sirius Shocker," *Billboard*, October 16, 2004, 73.

101. More on this in chapter 4.

102. Bill Werde, "SkyHigh: Music Business Seeks Its Own Windfall Amid Satellite Radio's Spending Spree," *Billboard*, November 26, 2005, 30–31.

103. Tamara Conniff, "Stern's Sirius Shocker," *Billboard*, October 16, 2004, 73.

104. Brian Garrity, "Media Industry Shares Gain on Wall Street," *Billboard*, January 15, 2005, 5, 48.

105. "Howard's End?," *Billboard*, April 8, 2006, 10.

106. Chris M. Walsh, "Poll: Sirius Closes Gap on XM," *Billboard*, June 10, 2006, 6. Survey participants included rock fans who paid for a Sirius subscription.

107. Chris M. Walsh, "The Latest News from .biz," *Billboard*, January 20, 2007, 7–8, 10.

108. Karmazin was Howard Stern's old boss at Viacom's Infinity Broadcasting and signed on as Sirius CEO around the time of the Stern deal.

109. Sirius Satellite Radio, *2004 Annual Report and Proxy Statement*, 2005.

110. Oliver Libaw, "Eminem on the Air," *Rolling Stone*, August 19, 2004, 34.

111. Sirius Satellite Radio, *2005 Annual Report and Proxy Statement*, 2006.

112. Bill Werde, "SkyHigh: Music Business Seeks Its Own Windfall Amid Satellite Radio's Spending Spree," *Billboard*, November 26, 2005, 30-31.

113. XM Satellite Radio Inc, *Form 10-K: Annual Report Pursuant to Section 13 or 15(d) of the Securities Exchange Act of 1934, for the Fiscal Year Ended December 31, 2005*, 2006, 22.

114. Ben Fritz, "Niche No More: XM, Sirius Pursue Talent as Radio Aud Grows," *Variety*, September 27, 2004, 4.

115. These costs include costs to create, produce, and acquire content, as well as on-air talent and broadcast royalties.

116. Sirius Satellite Radio, *2004 Annual Report and Proxy Statement*, 2005, 5, 7.

117. Sirius Satellite Radio, *2004 Annual Report and Proxy Statement*, 8.

118. Chris M. Walsh, "Sirius Losses, Subs Grow," *Billboard*, May 13, 2006, 6.

119. Paul Bond, "Satellite Radio in Limbo," *The Hollywood Reporter*, February 27, 2008, www.hollywoodreporter.com/business/business-news/satellite-radio-limbo-105748.

120. Sirius Satellite Radio, *2006 Annual Report and Proxy Statement*, 2007.

121. Paul Bond, "Satellite Radio in Limbo," *The Hollywood Reporter*, February 27, 2008, www.hollywoodreporter.com/business/business-news/satellite-radio-limbo-105748.

122. The implications of which are discussed in greater detail in chapter 4.

123. XM Satellite Radio, *2001 Annual Report*, 2002, 11.

124. Frank Saxe, "Satellite Radio to Hear Ads," *Billboard*, August 26, 2000, 89.

125. XM Satellite Radio, *2001 Annual Report*, 2002, 16.
126. Steven Graybrow, "Programming: Post-Teens Want Their New Music Too," *Billboard*, April 26, 2003, 21.
127 Graybrow, "Programming."
128. Sirius XM Radio Inc., *Proxy Statement and 2009 Annual Report*, 2010.
129. Car culture has been reflected in popular culture as aligning with freedom and rebelling, in films like *American Graffiti* and *Rebel Without a Cause*.
130. Bijsterveld et al., *Sound and Safe*, 4.
131. Russo, *Points on the Dial*, 171.
132. Sheller and Urry, "City and the Car," 747.
133. Baudrillard, *System of Objects*, 67.
134. Williams, *Television*, 18.
135. Bull, "Automobility and Power of Sound," 245.
136. Butch cited in Bull, "Automobility and Power of Sound," 245.
137. Bull, "Automobility and the Power of Sound," 247.
138. Bijsterveld et al., *Sound and Safe*, 2.
139. Featherstone, "Automobilities," 6.
140. Sheller and Urry, "New Mobilities Paradigm," 210–211.
141. Sheller and Urry, "New Mobilities Paradigm," 213.
142. Skeggs, *Class, Self, Culture*, 53.
143. Skeggs, *Class, Self, Culture*, 48.
144. Sterne, "Bourdieu, Technique and Technology," 375.
145. Peterson and Kern, "Changing Highbrow Taste"; and Webster, "Taste in the Platform Age," 1912.
146. "XM Takes Flight," *Billboard*, February 26, 2005, 48.
147. Sirius Satellite Radio Inc., *Form 10-K (Annual Report): Filed 3/12/2004 for Period Ending 12/31/2003*, 2004, 7.
148. XM Satellite Radio, *2001 Annual Report*, 2002, 16.
149. Michael Wolff, "Cruise Control," *New York Magazine*, September 25, 2000, https://nymag.com/nymetro/news/media/columns/medialife/3823.
150. XM Satellite Radio Holdings Inc., *Form 10-K (Annual Report): Filed 3/16/2000 for Period Ending 12/31/1999*, 2000, 12.
151. Katy Bachman, "NAB Calls for FCC Action against Satellite Radio," *Billboard*, September 8, 2001, 83.
152. Graham, *Vertical*, ix.
153. Bram Teitelman, "News Line . . . The Week in Brief," *Billboard*, August 14, 2004, 9.
154. Margo Whitmire, "You Can Call Jason 'Mr. A-Z,'" *Billboard*, July 23, 2005, 32; and Melinda Newman, "The Beat," *Billboard*, August 6, 2005, 38.
155. Brian Garrity, "Starbucks Spreading CD 'Bars,'" *Billboard*, October 23, 2004, 6, 75.
156. Lisa Gill, "Home Brew: Low Stars: Starbucks' First Previously Unreleased Act," *Billboard*, February 17, 2007, 34.

157. XM Satellite Radio Holdings Inc., *Form 10-K: Annual Report Pursuant to Section 13 or 15(d) of the Securities Exchange Act of 1934, for the Fiscal Year Ended December 31, 2007*, 2008, F-29.

158. Michael Paoletta, "Hotel Takes Bite of Apple," *Billboard*, March 5, 2005, 14.

159. Paul Heine, "XM Says Hi to Hyatt," *Billboard*, July 9, 2005, 9, 12.

160. Berland, "Locating Listening," 133.

161. XM Satellite Radio Inc., *Form 10-K: Annual Report Pursuant to Section 13 or 15(d) of the Securities Exchange Act of 1934, for the Fiscal Year Ended December 31, 2004*, 2005, 19.

162. XM Satellite Radio, *Form 10-K: Annual Report for December 31, 2004*, 19.

163. Styvén, "Intangibility of Music," 54.

164. Marc Schiffman, "Tuned In: Radio: AOL, XM Sail on a Stream," *Billboard*, April 23, 2005, 43.

165. Antony Bruno, "XM + Napster = New Download Option," *Billboard*, August 6, 2005, 11.

166. XM Satellite Radio Inc., *Form 10-K: Annual Report Pursuant to Section 13 or 15(d) of the Securities Exchange Act of 1934, for the Fiscal Year Ended December 31, 2005*, 2006, 5.

167. Antony Bruno, "Sprint Gets Sirius," *Billboard*, September 24, 2005, 8; and Chris M. Walsh, "The Latest News From .biz," *Billboard*, March 24, 2007, 5–6, 8.

168. Antony Bruno, "Digital Wrap-Up," *Billboard*, November 18, 2006, 29, 32.

169. Susan Butler, "Broadcast Flag Bill Flies into Controversy," *Billboard*, March 18, 2006, 13.

170. Berland, "Locating Listening," 132.

171. Streeter, "Blue Skies and Strange Bedfellows," 236.

3. PERCEIVED VALUE AND SATELLITE STREAMS IN MUSIC PROGRAMMING

1. Some services increase the quality with the amount one spends.

2. Eric Harvey, "Station to Station: The Past, Present, and Future of Streaming Music," *Pitchfork*, April 16, 2014, https://pitchfork.com/features/cover-story/reader/streaming.

3. Morris and Powers, "Control, Curation and Musical Experience," 107.

4. Spilker and Colbjørnsen, "Dimensions of Streaming."

5. "The XM Satellite Radio System," *Broadcast Engineering*, December 2002, 64.

6. Sirius Satellite Radio Inc., *Form 10-K: Annual Report Pursuant to Section 13 or 15(d) of the Securities Exchange Act of 1934, for Fiscal Year Ended December 31, 2002*, 2003, 3–4.

7. Sirius Satellite Radio, *Form 10-K: Annual Report for December 31, 2002*, 3–4.

8. XM Satellite Radio Holdings Inc., *Form 10-K: Annual Report Pursuant to Section 13 or 15(d) of the Securities Exchange Act of 1934, for the Fiscal Year Ended December 31, 2001*, 2002, 5.

9. XM Satellite Radio Inc., *Form 10-K: Annual Report Pursuant to Section 13 or 15(d) of the Securities Exchange Act of 1934, for the Fiscal Year Ended December 31, 2004*, 2005, 3.

10. XM Satellite Radio Holdings Inc., *Form 10-K: Annual Report Pursuant to Section 13 or 15(d) of the Securities Exchange Act of 1934, for the Fiscal Year Ended December 31, 2007*, 2008, 2.

11. XM Satellite Radio, *Form 10-K: Annual Report for December 31, 2007*, 4.

12. Sirius Satellite Radio Inc., *Form 10-K (Annual Report): Filed 02/29/08 for the Period Ending 12/31/07*, 2008, 8.

13. Now defunct but previously on channel 162; more on this channel in chapter 6.

14. These artist-based channels are described in more detail in chapter 5.

15. Chuck Taylor, "Satellite Service Could Signal Radio's Return to Creativity," *Billboard*, July 17, 1999, 102.

16. Taylor, "Radio's Return to Creativity," 102.

17. Lee Abrams, interview by Brian Fauteux, June 13, 2023.

18. Abrams interview.

19. Abrams interview.

20. Lee Abrams, "Random Thoughts, Clippings, Early XM and Sexism at CBS!," *Lee Abrams Media Visions*, June 18, 2007, www.leeabramsmediavisions.com/blog/random-thoughts-clippings-early-xm-and-sexism-at-cbs.

21. Abrams interview.

22. Anonymous, interview by Brian Fauteux, November 11, 2016.

23. Results determined by https://xmplaylist.com on May 8, 2024.

24. Anonymous interview.

25. Plasketes, "Romancing the Record," 109.

26. Marc Shiffman, "Sirius Launches New Streams," *Billboard*, January 31, 2004, 57.

27. "Introducing 'XM NATION,'" *Rolling Stone*, December 11, 2003, N3.

28. Jewell, *Live from the Underground*, 19.

29. Kramer, *Republic of Rock*, 69.

30. Gavin Edwards, "Satellite Radio Prepares for Liftoff," *Rolling Stone*, September 13, 2001, 101-102.

31. Bill Werde, "Satellite Radio Rocks Out," *Rolling Stone*, March 24, 2005, 12.

32. Bill Werde, "Rock Radio No Longer Rolling," *Rolling Stone*, March 24, 2005, 11–12.

33. Gavin Edwards, "Satellite Radio Prepares for Liftoff," *Rolling Stone*, September 13, 2001, 101–102.

34. Damien Cave, "Clear Channel: Inside Music's Superpower," *Rolling Stone*, September 2, 2004, 53–54.

35. Antony Bruno, "The Digital Future Is Not Yet Here," *Billboard*, June 18, 2005, 37, 40.

36. XM Satellite Radio Inc., *Form 10-K (Annual Report): Filed 4/2/2003 for Period Ending 12/31/2002*, 2003, 10.

37. Chris M. Walsh, "XM Embraces Reggaeton," *Billboard*, August 27, 2005, 10.

38. Leila Cobo, "Radio Finally Ready for Reggaton," *Billboard*, November 6, 2004, 33, 36.

39. Leila Cobo, "Latin Notas: BMG U.S. Latin Makes Cuts; Sirius Gets Serious," January 31, 2004, 25.

40. Phyllis Stark, "6 Questions with Eric Logan," *Billboard*, October 1, 2005, 58.

41. Fred Goodman, "Sales Up, Biz Flat in 2000," *Rolling Stone*, February 15, 2001, 21–22.

42. Williams, *Television*, 86, 91.

43. Nilay Patel, "SiriusXM's 360 Strategy, with CEO Jennifer Witz," *The Verge*, June 6, 2023, www.theverge.com/23750277/siriusxm-strategy-ceo-jennifer-witz-howard-stern-conan-obrien-radio-podcast-cars.

44. "National Satellite Report," *Arbitron*, Spring 2007, https://web.archive.org/web/20071030191216/http://www.allaccess.com/assets/editorial/raw/SP07_National_Satellite_P12.pdf.

45. Bürkner and Lange, "Sonic Capital," 34.

46. Stone, *Value of Popular Music*, xxix.

47. Regev, "Popular Music Studies," 22.

48. Frith, *Performing Rites*, 4.

49. Frith, *Performing Rites*, 18.

50. Frith, *Performing Rites*, 90.

51. Hesmondhalgh, *Why Music Matters*, 1–2.

52. Taylor, *Making Value*, 176–177.

53. Frith, *Performing Rites*, 273.

54. Frith, *Performing Rites*, 238

55. Gail Mitchell, "Rhythm, Rap, and the Blues," *Billboard*, August 10, 2002, 22.

56. "Buffett Debuts on Satellite Radio," *Rolling Stone*, June 2, 2005, 20.

57. Tamara Conniff, "Eminem's Back with a Vision," *Billboard*, November 20, 2004, 3, 91.

58. Oliver Libaw, "Eminem On the Air," *Rolling Stone*, August 19, 2004, 34.

59. David Wild, "The Ten Things That Piss Off Tom Petty," *Rolling Stone*, November 28, 2002, 34.

60. Mike Boyle, "'The Last DJ' Gets in the Last Word," *Billboard*, March 25, 2006, 54.

61. Craig Rosen, "Tom Petty & The Heartbreakers," *Billboard*, March 25, 2006, 35–38.

62. "Executive Action: Interscope/Geffen/A&M, Iovine Team with Sirius," *R&R*, January 28, 2005, 14.

63. Negus, "Plugging and Programming," 57.
64. Frith, "Look! Hear!," 278–279.
65. Tschmuck, *Creativity and Innovation*, 70.
66. Tschmuck, *Creativity and Innovation*, 78.
67. Frith, *Performing Rites*, 79.
68. Denisoff, *Tarnished Gold*.
69. Phyllis Stark, "6 Questions with Eric Logan," *Billboard*, October 1, 2005, 58.
70. "Inside XM," *Rolling Stone*, December 11, 2003, N21.
71. Marc Schiffman, "Getting to Know You: Labels Sniff Out Satellite Radio," *Billboard*, June 8, 2002, 68; and Harold Childs and Hilary Clay Hicks, "Radio, New-Media Promo Should Be On Same Track," *Billboard*, January 28, 2006, 4.
72. Simon Thompson, "Queen's 'Bohemian Rhapsody' Is Officially the World's Most-Streamed Song," *Forbes*, December 10, 2018, www.forbes.com/sites/simonthompson/2018/12/10/queens-bohemian-rhapsody-is-officially-the-worlds-most-streamed-song.
73. Daniel Kreps, "'Bohemian Rhapsody' Producer Lands Authorized Michael Jackson Biopic," *Rolling Stone*, November 22, 2019, www.rollingstone.com/music/music-news/bohemian-rhapsody-producer-michael-jackson-biopic-916842.
74. Yohana Desta, "Let's Go Crazy: An Original Prince Movie, Inspired by His Music, Is in the Works," *Vanity Fair*, December 3, 2018, www.vanityfair.com/hollywood/2018/12/prince-original-movie-music; and Etan Vlessing, "Prince Music-Inspired Movie in the Works at Universal," *The Hollywood Reporter*, December 3, 2018, www.hollywoodreporter.com/news/prince-music-inspired-movie-works-at-universal.
75. Jack Beresford, "U2 Radio Station Playing Nothing but Songs from the Irish Music Icons to Launch in 2020," *The Irish Post*, December 5, 2019, www.irishpost.com/news/u2-radio-station-174807.
76. Gil Kaufman, "The Edge Says U2's SiriusXM Channel Has 'Power to Engage in an Emotional and Deep Way,'" *Billboard*, June 30, 2020, www.billboard.com/articles/columns/rock/9409747/the-edge-interview-u2-siriusxm-xradio.
77. Carla Hay, "Duran Duran Gets Sirius with Elvis," *Billboard*, October 23, 2004, 49.
78. DeNora, *Music in Everyday Life*, 61.
79. Rodman, "Radio Formats in the United States," 235.
80. Berland, "Radio Space and Industrial Time," 181.
81. Roessner, "Radio in Transit," 56.
82. Roessner, "Radio in Transit," 61.
83. Alper, "XM Reinvents Radio's Future," 515.
84. Steven Graybrow, "Radio Stations Hit the Target Audience," *Billboard*, May 31, 2003, 49, 55.
85. Weisbard, *Top 40 Democracy*, 72.

86. Watson, "Gender on Hot Country Chart."
87. Lee Abrams, "Random Stuff from the XM Halls," *Lee Abrams Media Visions*, August 21, 2006, www.leeabramsmediavisions.com/blog/random-stuff-from-the-xm-halls.
88. Negus, "Plugging and Programming," 63.
89. Coddington, *Hip Hop became Hit Pop*, 1.
90. Elias Leight, "Top 40 Radio Has a Rap Problem," *Rolling Stone*, May 31 2018, www.rollingstone.com/music/music-news/top-40-radio-has-a-rap-problem-630658.
91. Rashaun Hall and Gail Mitchell, "Satellite Radio Gives Lyrics Free Rein," *Billboard*, December 20, 2003, 50.
92. Allison Stewart, "What Genres Have Benefited the Most from the Streaming Era of Music?," *Washington Post*, April 3, 2018, www.washingtonpost.com/lifestyle/what-genres-have-benefited-the-most-from-the-streaming-era-of-music/2018/04/03/950e99a8-3695-11e8-9c0a-85d477d9a226_story.html; and Grant Rindner, "How Streaming Changed Rap," *Complex*, November 19 2019, www.complex.com/music/how-music-streaming-changed-rap.
93. XM Satellite Radio Inc., *Form 10-K405: Annual Report (Regulation S-K, item 405): Filed 3/19/2002 for Period Ending 12/31/2001*, 2002, 4.
94. XM Satellite Radio Inc., *Form 10-K: Annual Report Pursuant to Section 13 or 15(d) of the Securities Exchange Act of 1934, for the Fiscal Year Ended December 31, 2005*, 2006, 9.
95. XM Satellite Radio Holdings Inc., *Form 10-K (Annual Report): Filed 3/16/2000 for Period Ending 12/31/1999*, 2000, 3.
96. Ken Schlager, "NY'C' Is for Country," *Billboard*, November 19, 2005, 4.
97. Edward Morris, "Indie Artists Mostly Ignored by Radio," *Billboard*, March 18, 1995, 31.
98. Phyllis Stark, "Nashville Scene: CRS Tackles How to Revitalize Country Format," *Billboard*, June 12, 2004, 31.
99. Deborah Evans Price, "Blue Skies for Bluegrass," *Billboard*, October 9, 2004, 33, 35.
100. Deborah Evans Price and Phyllis Stark, "Bluegrass Is Growing," *Billboard*, October 29, 2005, 62–63. As of March 31, 2022, the FCC reported a total of 15,390 stations in the United States. See "Broadcast Station Totals as of March 31, 2022," *FCC*, www.fcc.gov/document/broadcast-station-totals-march-31-2022.
101. Paul Heine, "Satellite Radio Nabs Outlaw Country Acts," *Billboard*, March 5, 2005, 5, 73.
102. Chuck Taylor. "Summit Examines Dance Music's Status," *Billboard*, July 31, 1999, 3, 104.
103. Wodke, "Irony and the Ecstasy."
104. Chuck Taylor, "Dance Sees the Future Online," *Billboard*, July 31, 1999, 1, 104.

105. Michael Paoletta, "Coates Gives Voice to C2/Columbia's Madison Avenue," *Billboard*, October 28, 2000, 35.

106. Michael Paoletta, "Up Is Down for Remix Business," *Billboard*, August 30, 2003, 5, 65.

107. Barnett, *Record Cultures*, 109.

108. McLeod, "Space Oddities," 349.

109. Reynolds, *Retromania*, 368.

110. Reynolds, *Retromania*, xix.

111. Reynolds, *Retromania*, 160.

112. Van der Hoven, "Popular Music of the 1990s"; and Kotarba, "Music as a Timepiece."

113. XM Satellite Radio Inc., *Form 10-K (Annual Report): Filed 4/2/2003 for Period Ending 12/31/2002*, 2003, 9–10.

114. Named after the Nirvana song.

115. "Was 'Blog Radio: The Carles Show' Cancelled?," *reddit*, October 7, 2022, www.reddit.com/r/siriusxm/comments/xusx6d/was_blog_radio_the_carles_show_cancelled.

116. Lee Abrams, "Willie Nelson Travelogue and XM Thinking from the Early Days," *Lee Abrams Media Visions*, April 9, 2007, www.leeabramsmediavisions.com/blog/willie-nelson-travelogue-and-xm-thinking-from-the-early-days.

117. Bull, "Auditory Nostalgia of iPod Culture," 85; and Bijsterveld and Jacobs, introduction, 12

118. McCourt, "Music in the Digital Realm," 250.

119. Baudrillard, *Simulations*, 12.

120. Deleuze, "Plato and the Simulacrum," 49.

121. Williams, *Long Revolution*.

122. Jameson, *Postmodernism*, xx.

123 Jameson, *Postmodernism*, 4–5.

124. Jameson, *Postmodernism*, 17–18.

125. Fisher, *K-Punk*, 585.

126. XM Satellite Radio Inc., *Form 10-K: Annual Report Pursuant to Section 13 or 15(d) of the Securities Exchange Act of 1934, for the Fiscal Year Ended December 31, 2005*, 2006, 1.

127. Phyllis Stark, "Chicks Show Why They Rule the Roost; Hall of Fame, Satellite Broadcaster Link Up," *Billboard*, August 5, 2000, 37.

128. Bennett, "'Heritage Rock.'"

129. Gibson and Connell, "Music, Tourism and Memphis."

130. Baker and Huber, "Saving 'Rubbish,'" 117.

131. Phyllis Stark, "Nashville Scene: AMA Names Green Executive Director," *Billboard*, August 21, 2004, 33.

132. Phyllis Stark, "Nashville Scene: Sirius, XM Beam in on Nashville," *Billboard*, August 7, 2004, 37.

133. Phyllis Stark, "Nashville Scene: Country Stars Take DIY Route for New Albums," *Billboard*, August 12, 2006, 73.
134. Ken Tucker, "The Billboard Green 10," *Billboard*, March 29, 2008, 17.
135. Dan Ouellette, "Jazz Notes," *Billboard*, June 25, 2005, 46.
136. Yochim and Biddinger, "'That Vintage Feel,'" 183.
137. Fisher, *K-Punk*, 319.
138. Shumway, "Production of Nostalgia," 39.
139. Holbrook and Schindler, "Exploratory Findings on Musical Tastes," 119.
140. Lee Abrams, interview by Brian Fauteux, June 13, 2023.
141. Beaster-Jones, *Music Commodities, Markets, and Values*, 1.
142. Jesus Chigley, "Metric: Old World Underground, Where Are You Now?," *Drowned in Sound*, May 30, 2005, https://drownedinsound.com/releases/4098/reviews/12391-metric-old-world-underground-where-are-you-now.
143. "Albums: Old World Underground, Where Are You Now," *Pitchfork*, September 24, 2003, https://pitchfork.com/reviews/albums/5547-old-world-underground-where-are-you-now.
144. Tristan Young, "Looking Back on Metric's Debut Album, Old World Underground Where Are You Now?," *Midrange Weekly*, January 15, 2021, www.midrangevancouver.com/posts/2021/1/14/looking-back-on-metrics-debut-album-old-world-underground-where-are-you-now.
145. Taylor, "Performance and Nostalgia," 98–99.
146. Van der Hoeven, "Popular Music of the 1990s," 320; and van der Hoeven, "Songs that Resonate."
147. Camille, "Simulacrum," 38.
148. Tacchi, "Nostalgia and Radio Sound," 282.
149. Tacchi, "Nostalgia and Radio Sound," 281.
150. Livingston, "Music Revivals," 66; and Shumway, "Production of Nostalgia," 37.
151. Plasketes, "Romancing the Record," 117–118.

4. DISPOSABLE MUSIC AND MONOPOLY POWER IN POLICYMAKING

1. Marcel Boyer, *C. D. Howe Institute, Commentary No. 419: The Value of Copyrights in Recorded Music: Terrestrial Radio and Beyond*, February 2015; and Frances W. Preston, "President's Letter: A New World Order in Copyright Protection," *BMI*, Winter 1993, 2.
2. Omer Anderson, "Melody Protection Stressed," *Billboard*, June 27, 1964, 10.
3. Brian Hiatt and Evan Serpick, "The Record Industry's Slow Fade," *Rolling Stone*, June 28, 2007, 13–14.
4. Felix Richter, "From Tape to Tidal: 4 Decades of U.S. Music Sales," *Statista*, June 24, 2022, www.statista.com/chart/17244/us-music-revenue-by-format; and

NOTES 257

David Goldman, "Music's Lost Decade: Sales Cut in Half," *CNN Money*, February 3, 2010, https://money.cnn.com/2010/02/02/news/companies/napster_music_industry.

5. Georg Szalai, "Sirius XM Radio Ended 2010 with More Subscribers Than Netflix," *The Hollywood Reporter*, February 15, 2011, www.hollywoodreporter.com/business/business-news/sirius-xm-radio-ended-2010-99625.

6. "'Streaming Grossly Underpays Writers,'" *Music Week*, June 14, 2013, 16.

7. Yearly earnings, however, were still down due to high expenses and paying down debt.

8. "The Action," *Billboard*, February 15, 2014, 6.

9. Andrew Dansby et al., "Random News," *Rolling Stone*, March 6, 2003, 22.

10. Murdock, "Back to Work," 32.

11. Susan Butler, "Legal Matters: The Urge to Merge," *Billboard*, March 10, 2007, 19.

12. Colby Cosh, "A Marriage Made in Outer Space," *National Post*, February 23, 2007, A14.

13. "Merchants & Marketing: In the News," *Billboard*, June 1, 2002, 62.

14. Matthew Benz and Erik Gruenwedel, "Satellite Radio Debate among Investors, Analysts Rumbles On," *Billboard*, August 17, 2002, 33.

15. Brian Garrity, "Music & Money," *Billboard*, November 30, 2002, 56.

16. Matthew Benz, "XM, Sirius Grow Subs, Losses," *Billboard*, April 12, 2003, 6.

17. Matthew Benz, "Music-Related Stocks Enjoy Strong First Half," *Billboard*, July 19, 2003, 5.

18. Andy Kessler, "Satellite Radio Gets Sirius," *Wall Street Journal*, March 8, 2005, B2.

19. Brian Garrity, "And Our Money's On . . . ," *Billboard*, January 6, 2007, 36.

20. Jeffrey Yorke, "Still Orbiting," *Billboard*, September 22, 2007, 16.

21. *Hearing Before the Antitrust Task Force of the House Committee on the Judiciary*, 110th Cong. 1 (February 28, 2007).

22 *Hearing Before the Antitrust Task Force*, 83.

23. *Hearing Before the Antitrust Task Force*, 31.

24. *Hearing Before the Antitrust Task Force*, 2.

25. *Hearing Before the Antitrust Task Force*, 49.

26. Chris M. Walsh, "The Latest News from .biz, Satcasters to Offer a la Carte Programming," *Billboard*, August 4, 2007, 8.

27. Jeffrey Yorke, "Still Orbiting," *Billboard*, September 22, 2007, 16.

28. Yorke, "Still Orbiting," 16.

29. Holt, *Empires of Entertainment*, 3.

30. Chris M. Walsh, "The Latest News from .biz. Sat Merger Likely to Be Approved," *Billboard*, August 2, 2008, 8.

31. Paul Bond, "Tate's Yes Vote Will Seal Sat Deal," *The Hollywood Reporter*, July 24, 2008, www.hollywoodreporter.com/business/business-news/tates-yes-vote-will-seal-116233.

32. Some other figures had estimated $5 billion in total. Kim Hart, "Satellite Radio Merger Approved," *Washington Post*, July 26, 2008, www.washingtonpost.com/wp-dyn/content/article/2008/07/25/AR2008072503026.html?hpid=topnews; and Brooks Boliek, "FCC Now Final Hurdle for Satellite Radio Merger," *The Hollywood Reporter*, March 25, 2008, www.hollywoodreporter.com/business/business-news/fcc-final-hurdle-satellite-radio-107771.

33. Federal Communications Commission, *Commission Approves Transaction between Sirius Satellite Radio Holdings Inc. and XM Satellite Radio Holdings, Inc. Subject to Conditions* (Washington, DC: FCC, 2008).

34. *In the Matter of Determination of Rates and Terms for Digital Performance in Sound Recordings and Ephemeral Recordings (Web IV), Before the United States Copyright Royalty Judges, The Library of Congress, Washington, DC*, Docket No. 14-CRB-0001-WR (October 6, 2014), 3, 16 (written direct testimony of David J. Frear on behalf of Sirius XM Radio Inc.).

35. XM Satellite Radio Holdings Inc., *Form 10-K: Annual Report Pursuant to Section 13 or 15(d) of the Securities Exchange Act of 1934, for the Fiscal Year Ended December 31, 2008*, 2009, 9.

36. XM Satellite Radio Holdings Inc., *Form 10-K: Annual Report Pursuant to Section 13 or 15(d) of the Securities Exchange Act of 1934, for the Fiscal Year Ended December 31, 2006*, 2007, 16.

37. Sirius Satellite Radio Inc., *Form 10-K (Annual Report): Filed 02/29/08 for the Period Ending 12/31/07*, 2008, 13

38. Sirius XM Radio Inc., *Form 10-K (Annual Report): Filed 03/10/09 for the Period Ending 12/31/08*, 2009, 14.

39. XM Satellite Radio Holdings Inc., *Form 10-K: Annual Report Pursuant to Section 13 or 15(d) of the Securities Exchange Act of 1934, for the Fiscal Year Ended December 31, 2006*, 15.

40. Sirius Satellite Radio Inc., *Form 10-K (Annual Report): Filed 02/29/08 for the Period Ending 12/31/07*, 2008, 4.

41. Darlene Darcy, "Severance Deals Protect XM Satellite Executives," *Washington Business Journal*, August 4, 2008.

42. Sirius XM Radio Inc., *2007 Annual Report and Proxy Statement*, 2008, 1.

43. Sirius XM Radio Inc., *Proxy Statement and 2009 Annual Report*, 2010.

44. Sirius XM Radio Inc., *Form 10-K (Annual Report): Filed 02/25/10 for the Period Ending 12/31/09*, 2010, 42.

45. Lee Abrams, interview by Brian Fauteux, June 13, 2023.

46. Ken Tucker, "Station Break," *Billboard*, April 26, 2008, 5.

47. Matt Hengeveld, "SiriusXM's 'The Village' Now Web-Only," *Sing Out!*, January 10, 2012, https://singout.org/siriusxms-the-village-now-web-only.

48. Antony Bruno, "Sirius Problems," *Billboard*, November 15, 2008, 17.

49. Paul Bond, "Shock Options: Sirius XM Star Seeks New Pact as Fiscal Static Plagues Radio Biz," *The Hollywood Reporter*, February 4, 2010, 1, 14.

NOTES   259

50. Chris M. Walsh, "The Latest News from .biz: Reports, Sirius XM in Talks with DirecTV, Liberty," *Billboard*, February 21, 2009, 6.

51. Chris M. Walsh, "The Latest News from .biz: Sirius XM Agrees to Liberty Deal," *Billboard*, February 28, 2009, 5.

52. Paul Bond and Georg Szalai, "Liberty Call Saves Sat Radio from Bankruptcy," *The Hollywood Reporter*, February 18, 2009, www.hollywoodreporter.com/business/business-news/liberty-call-saves-sat-radio-79462.

53. Sirius XM Radio Inc., *Proxy Statement and 2009 Annual Report*, 2010, 21.

54. Ed Christman, "Scott Greenstein," *Billboard*, July 26, 2014, 20.

55. Sirius XM Satellite Radio, *Proxy Statement and 2012 Annual Report*, 2013.

56. Sirius XM Holdings Inc., *Form 10-K (Annual Report): Filed 02/04/14 for the Period Ending 12/31/13*, 2014, 3.

57. Baker and Huber, "Saving 'Rubbish,'" 113.

58. Taylor, "'I Am What I Play,'" 82.

59. Jones, "Music Industry as Workplace," 149.

60. Toynbee, "Fingers to the Bone," 40.

61. Jones, "Music Industry as Workplace," 155.

62. Bill Werde with Susan Butler and Tony Sanders, "Sky High: Music Business Seeks Its Own Windfall Amid Satellite Radio's Spending Spree," *Billboard*, November 26, 2005, 31.

63. Antony Bruno, "Is Time-Shifting Downloading?," *Billboard*, October 29, 2005, 15.

64. Bill Werde with Susan Butler and Tony Sanders, "Sky High: Music Business Seeks Its Own Windfall Amid Satellite Radio's Spending Spree," *Billboard*, November 26, 2005, 31.

65. Gibson, "Recording Studios," 196.

66. "Artists Withdraw from Spotify, Dis Pandora," *International Musician*, August 1, 2013, 6.

67. Stuart Dredge, "Daniel Ek: Spotify and Free Music Will Save the Industry, Not Kill It," *The Observer*, June 7, 2015, www.theguardian.com/technology/2015/jun/07/daniel-ek-spotify-free-music-save-industry-not-kill-it.

68. Eriksson et al., *Spotify Teardown*, 50.

69. Eriksson et al., *Spotify Teardown*, 64.

70. Alan B. Krueger and Ying Zhen, "MIRA Survey of Musicians," *Music Industry Research Association*, June 19, 2018, https://psrc.princeton.edu/news/mira-survey-musicians-april-june-2018.

71. Klein, Meier, and Powers, "Selling Out."

72. Meier, *Popular Music as Promotion*, 153, 160.

73. Meier, *Popular Music as Promotion*, 3.

74. *In the Matter of Determination of Rates and Terms for Digital Performance in Sound Recordings and Ephemeral Recordings (Web IV), Before the United States Copyright Royalty Judges, The Library of Congress, Washington, DC,*

260   NOTES

Docket No. 14-CRB-0001-WR (October 6, 2014), 3 (written direct testimony of David J. Frear on behalf of Sirius XM Radio Inc.).

75. *In the Matter of Determination of Rates and Terms*, 4.

76. *In the Matter of Determination of Rates and Terms*, 5.

77. Sirius Satellite Radio Inc., *Form 10-K: Annual Report Pursuant to Section 13 or 15(d) of the Securities Exchange Act of 1934, for Fiscal Year Ended December 31, 2002*, 2003, 4.

78. In most other countries with established radio and record industries, copyright laws pay performers as well as the songwriters and publishers for the public performance of music. See Susan Butler, "UpFront: Legal—Will Radio Pay Artists and Labels?," *Billboard*, August 11, 2007, 6.

79. Kronenberg, "Royalty Rates and Exclusive Releases."

80. Kronenberg, "Royalty Rates and Exclusive Releases," 636.

81. Chuck Taylor, "Broadcasters Sue RIAA over Royalties," *Billboard*, April 8, 2000, 76.

82. Ray Hair, "Big Radio: Deal with Us on Performance Rights Before It's Too Late," *International Musician*, December 28, 2017, https://internationalmusician.org/big-radio. Also, broadcasters do indeed pay royalties to SoundExchange for simulcast streaming and have fought against these rates, claiming they are too high.

83. Bill Holland, "The Last Word: A Q&A with Ann Chaitovitz," *Billboard*, April 16, 2005, 74.

84. Susan Butler, "UpFront: Legal—Will Radio Pay Artists and Labels?," *Billboard*, August 11, 2007, 6.

85. Kidd, "Internet Radio Royalty Dispute," 15.

86. Frank Saxe, "Radio, Record Labels Chafe Over Streaming," *Billboard*, May 26, 2001, 1.

87. Ed Christman, "Rethinking Radio," *Billboard*, June 16, 2012, 8.

88. Glenn Peoples, "Major Radio Play," *Billboard*, September 21, 2013, 4.

89. Susan Butler, "News Line: The Week in Brief," *Billboard*, August 21, 2004, 10.

90. Bill Holland, "Digital $$'s Stream In," *Billboard*, December 4, 2004, 1, 55.

91. Kronenberg, "Royalty Rates and Exclusive Releases," 633, 634.

92. Kronenberg, "Royalty Rates and Exclusive Releases," 638.

93. Susan Butler, "News Line: The Week in Brief," *Billboard*, August 21, 2004, 10.

94. Bill Holland, "The Last Word: A Q&A with Ann Chaitovitz," *Billboard*, April 16, 2005, 74.

95. Bill Holland, "Digital $$'s Stream In," *Billboard*, December 4, 2004, 1, 55.

96. Bill Holland, "The Last Word: A Q&A with Ann Chaitovitz," *Billboard*, April 16, 2005, 74.

97. Glenn Peoples, "Digital Music's Rise," *Billboard*, October 26, 2013, 8.

98. Anna Washenko, "SoundExchange Posts Record Distribution Payouts in 2018," *RAIN News*, January 31, 2019, https://rainnews.com/soundexchange-posts-record-distribution-payouts-in-2018.

99. Anna Washenko, "Sirius XM Posts Record High Revenue in 2018," *RAIN News*, January 31, 2019, https://rainnews.com/sirius-xm-posts-record-high-revenue-in-2018.

100. Brian Garrity, "New Revenue On Tap," *Billboard*, April 15, 2006, 30.

101. XM Satellite Radio Holdings Inc., *Form 10-K: Annual Report Pursuant to Section 13 or 15(d) of the Securities Exchange Act of 1934, for the Fiscal Year Ended December 31, 2007*, 2008, 26.

102. Sirius XM Radio Inc., *2007 Annual Report and Proxy Statement*, 2008, 12.

103. Antony Bruno and Ed Christman, "The Latest News from .biz: Sirius XM Seeks Direct Label Licensing," *Billboard*, August 20, 2011, 4.

104. Ed Christman, "Action Pact," *Billboard*, September 10, 2011, 5.

105. Ed Christman, "Rethinking Radio," *Billboard*, June 16, 2012, 8.

106. See Future of Music Coalition (@future_of_music), "Some folks wonder why satellite radio pays more per spin than subscription interactive streaming services do. Let's walk through the math," Twitter, August 5, 2022, https://twitter.com/future_of_music/status/1555558808637808645.

107. John Villasenor, "The Satellite Question," *Billboard*, March 2, 2013, 13.

108. Sirius XM Radio Inc., *Form 10-K (Annual Report): Filed 02/09/12 for the Period Ending 12/31/11*, 2012, 9.

109. John Villasenor, "The Satellite Question," *Billboard*, March 2, 2013, 13.

110. Sirius XM Holdings Inc., *Form 10-K: Annual Report Pursuant to Section 13 or 15(d) of the Securities Exchange Act of 1934 for the Fiscal Year Ended December 31, 2017*, 2018, 9.

111. Ed Christman, "Copyright Royalty Board Raises Rate for SiriusXM, Lowers It for Music Choice," *Billboard*, December 15, 2017, www.billboard.com/articles/business/8070762/sirius-xm-copyright-royalty-board-crb-rate-increase.

112. Christman, "Board Raises Rate for SiriusXM.".

113. Ed Christman, "SiriusXM's Copyright Defeat," *Billboard*, October 4, 2014, 12.

114. Ed Christman, "Turtle Power," *Billboard*, August 17, 2013, 4

115. Ed Christman, "SiriusXM's Copyright Defeat," *Billboard*, October 4, 2014, 12.

116. Sirius XM Holdings Inc., *Form 10-K (Annual Report): Filed 02/05/15 for the Period Ending 12/31/14*, 2015, 12.

117. Ed Christman, "SiriusXM's $40 Million Lose-Win," *Billboard*, December 10, 2016, 24.

118. Jonathan Stempel, "Sirius XM Wins Dismissal of Turtles Copyright Lawsuit in New York," *Reuters*, February 16, 2017, www.reuters.com/article/us-sirius-xm-holdgs-ruling-turtles-idUSKBN15V27U.

119. Eriq Gardner, "Did SiriusXM Pull a Fast One on Major Record Labels in a Deal with Suing Indie Musicians?," *The Hollywood Reporter*, www.hollywoodreporter.com/thr-esq/did-siriusxm-pull-a-fast-one-major-record-labels-a-deal-suing-indie-musicians-992733.

120. Flo & Eddie, Inc., a California Corporation, Individually and on Behalf of All Others Similarly Situated, Plaintiffs, v. Sirius XM Radio Inc., a Delaware Corporation, and DOES 1 through 10, Defendants, Case No. CV 13-05693 PSG (GJS), U.S. Dist. Ct. Cent. Dist. Cal. (May 1, 2017). (Defendant Sirius XM Radio Inc.'s Opposition to Amici Curiae's request for leave to file brief regarding class settlement [Declaration of Vision Winter filed concurrently herewith]).

121. Steve Knopper, "Music's New Best Frenemy: SiriusXM," *Billboard*, July 28, 2017, 14.

122. Robert Levine, "Backstage Pass: Music and Power on the Potomac," *Billboard*, April 14, 2018, 51–53.

123. *Hearing on Protecting and Promoting Music Creation for the 21st Century, Before the Senate Judiciary Committee*, 115th Cong. (May 15, 2018) (statement by Senator Chuck Grassley of Iowa, chairman).

124. Robert Levine, "Backstage Pass: Music and Power on the Potomac," *Billboard*, April 14, 2018, 51–53.

125. Melinda Newman, "Can the Music Modernization Act Beat the Clock?," *Billboard*, August 11, 2018, 26.

126. Music Modernization Act, S. Rep. No. 115-339 (September 17, 2018), www.govinfo.gov/content/pkg/CRPT-115srpt339/html/CRPT-115srpt339.htm.

127. *Hearing on Protecting and Promoting Music Creation for the 21st Century, Before the Senate Committee on the Judiciary*, 115th Cong. 2 (May 15, 2018) (written testimony of David J. Del Beccaro, president and CEO of Music Choice).

128. *Hearing on Protecting and Promoting Music Creation for the 21st Century, Before the Senate Committee on the Judiciary*, 115th Cong. 5 (May 15, 2018) (responses of David J. Del Beccaro to Senator Grassley's written questions).

129. *Hearing on S. 2823, the Music Modernization Act, Before the Senate Committee on the Judiciary*, 115th Cong. 1 (May 15, 2018) (statement of Mitch Glazier, president of Recording Industry Association of America).

130. *Hearing on Protecting and Promoting Music Creation for the 21st Century, Before the Senate Committee on the Judiciary*, 115th Cong. 1 (May 15, 2018) (statement of Christopher Harrison, chief executive officer, Digital Media Association [DiMA]).

131. *Hearing on Protecting and Promoting Music Creation for the 21st Century, Before the Senate Committee on the Judiciary*, 115th Cong. 5 (May 15, 2018) (statement of Smokey Robinson).

132. Ted Johnson, "Senators Introduce Bill to Extend Copyright to Classic Recordings," *Variety*, February 7, 2018, http://variety.com/2018/politics/news/sound-recordings-1972-pandora-siriusxm-1202691153.

133. Melinda Newman, "Women in Music 2018 Executives of the Year," *Billboard*, December 6, 2018, www.billboard.com/music/awards/billboard-women-in-music-executives-of-the-year-2018-8488600.

134. Dina LaPolt, "Why Live Nation's Chairman of the Board Should Resign," *Variety*, September 14, 2018, https://variety.com/2018/music/opinion/op-ed-music-modernization-act-live-nation-siriusxm-dina-lapolt-1202942884.

135. Melinda Newman, "Women in Music 2018 Executives of the Year," *Billboard*, December 6, 2018, www.billboard.com/music/awards/billboard-women-in-music-executives-of-the-year-2018-8488600.

136. Steve Knopper, "Will Labels Have a Sirius Problem?," *Billboard*, February 9, 2019, 22.

137. Anna Washenko, "Music Modernization Act for Mechanical Licensing Overhaul Leaps Major Hurdle with Unanimous Senate Approval," *RAIN News*, September 19, 2018, https://rainnews.com/music-modernization-act-for-mechanical-licensing-overhaul-leaps-major-hurdle-with-unanimous-senate-approval.

138. Brad Hill, "SiriusXM Earnings Call: Remarks on the Pandora Acquisition," *RAIN News*, October 26, 2018, https://rainnews.com/sirius-xm-earnings-call-remarks-on-the-pandora-acquisition.

139. "Frequently Asked Questions," *Copright.gov*, https://www.copyright.gov/music-modernization/faq.html.

140. Ed Christman, "SoundExchange Could See Collected Revenue Shrink by $200 Million in 2017: Exclusive," *Billboard*, December 29, 2016, www.billboard.com/music/music-news/soundexchange-could-see-revenue-shrink-200-million-2017-exclusive-7640450.

141. DeWaard, Fauteux, and Selman, "Independent Canadian Music."

142. Tom Pakinkis, "Streaming Services 'Have All Ignored Artist Needs,'" *Music Week*, April 11, 2014, 2.

143. Steve Knopper, "YouTube vs. the Music Industry," *Rolling Stone*, August 25, 2016, 22–23.

144. Justin Wm. Moyer, "Spotify Sued over Royalties," *Los Angeles Times*, December 31, 2015, C3.

5. THE STARS DOWN TO EARTH

1. "Beatles' Cut to Be Seen on Global TV," *Billboard*, May 27, 1967, 4.

2. Gillian G. Gaar, "Pearl Jam Hack the Airwaves," *Rolling Stone*, February 23, 1995, www.rollingstone.com/music/music-news/pearl-jam-hack-the-airwaves-180898.

3. Harold Childs and Hilary Clay Hicks, "Our View: The DJ Connection: Personality Radio Is Missing Link in Marketing Chain," *Billboard*, March 12, 2005, 10.

4. For other examples, see Razlogova, "The Past and Future of Music Listening"; and Cwynar, "Brick, Mortar, and Screen."
5. Johnson, "Date with the Duke," 369, 371.
6. Glenn Peoples, "Music Subscription Battle Gets Real," *Billboard*, January 26, 2013, 7.
7. Scannell, *Radio, Television and Modern Life*, 23.
8. Bottomley, *Sound Streams*, 15.
9. Peters, *Speaking into the Air*, 215.
10. Frith, *Performing Rites*, 186.
11. Lacey, "Up in the Air?," 121.
12. Priestman, "Radio's Great Global Conversation," 80, 84.
13. Priestman, "Radio's Great Global Conversation," 84.
14. Priestman, "Radio's Great Global Conversation," 84.
15. Roessner, "Radio in Transit," 65–66.
16. Shirley Halperin, "Who Plays DJ in the Digital Age?," *Hollywood Reporter*, June 7, 2013, 34; and Stiernstedt, "Political Economy of the Radio Personality," 290.
17. Reynolds, *Retromania*, 129.
18. Lee Abrams, interview by Brian Fauteux, June 13, 2023.
19. Marci Kenon, "You've Come a Long Way, Baby," *Billboard*, August 19, 2000, 46.
20. Jim Bessman, "Legendary Bill Anderson Chats with Country Icons on XM," *Billboard*, November 30, 2002, 28; and Moira McCormick, "What's Up on the Airwaves," *Billboard*, April 5, 2003, 24.
21. Gail Mitchell, "Flexing His Branding Muscle, Snoop Became a Household Name," *Billboard*, July 12, 2003, 18–20.
22. XM Satellite Radio, *2004 Annual Report*, 2005.
23. Lee Abrams, interview by Brian Fauteux, June 13, 2023.
24. Carla Hay, "Sirius Star Magnet," *Billboard*, January 31, 2004, 41.
25. Sirius Satellite Radio Inc., *Form 10-K (Annual Report): Filed 02/29/08 for the Period Ending 12/31/07*, 2008, 5.
26. Mike Stern, "The Power 100: Scott Greenstein," *Billboard*, February 16, 2013, 56.
27. Ed Christman, "Scott Greenstein: The Satellite Vet on Firing an Opie & Anthony Host, Adding EDM Channels and Sharing Data," *Billboard*, July 26, 2014, 20.
28. Lee Abrams, "The XM Radio Theater: Bob Dylan, George Carlin And Larry King," *Lee Abrams Media Divisions*, April 30, 2007, www.leeabramsmediavisions.com/blog/the-xm-radio-theater-bob-dylan-george-carlin-and-larry-king.
29. "Welcome to the Theme Time Radio Hour Archive," *Theme Time Radio Hour Archive*, www.themetimeradio.com/welcome-to-the-theme-time-radio-hour-archive.
30. Negus, *Bob Dylan*, 70.

31. Motoko Rich, "Who's This Guy Dylan Who's Borrowing Lines from Henry Timrod?," *New York Times*, September 14, 2006, www.nytimes.com/2006/09/14/arts/music/14dyla.html; and Scobie, "Plagiarism, Bob, Jean-Luc and Me."

32. Randy Lewis, "Is This A Real Case of Love and Theft?," *Los Angeles Times*, July 9, 2003, E1.

33. Robert Hilburn, "Rock's Enigmatic Poet Opens a Long-Private Door," *Los Angeles Times*, April 4 2004, A1.

34. Marshall, *Bob Dylan*, 76.

35. Moore, "Authenticity as Authentication," 211.

36. Svec, *American Folk Media*, 13.

37. Marshall, *Bob Dylan*, 78.

38. See also Keightley, "Reconsidering Rock."

39. Jon Pareles, "The Pilgrim's Progress of Bob Dylan," *New York Times*, August 20, 2006, A21.

40. Elliott, "Same Distant Places," 250.

41. Elliott, "Same Distant Places," 250.

42. Child, "Raised in the Country," 199–200.

43. In this way, Dylan's voice connects with the discussion of audio ambiance channels in chapter 3.

44. "Cars," *Theme Time Radio Hour*, July 19, 2006.

45. O'Malley, "Freeform Radio," 172.

46. Dylan, *Chronicles*, 188.

47. Some readers may be familiar with the character Davis on HBO's *Treme*, who is regularly scolded by his superiors for playing too much bounce music on his WWOZ radio program.

48. "Friends & Neighbors," *Theme Time Radio Hour*, August 23, 2006.

49. "Radio," *Theme Time Radio Hour*, 2006.

50. Caldwell, "Second Shift Media Aesthetics," 131.

51. Bolter and Grusin, *Remediation*, 14-15.

52. Moscote Freire, "Remediating Radio," 102.

53. Dylan, *Chronicles*, 32.

54. Douglas, *Listening In*, 3-4.

55. Joe Levy, "Modern Times," *Rolling Stone*, September 7 2006, 99.

56. "Baseball," *Theme Time Radio Hour*, May 24, 2006.

57. Bob Dylan, "It's All Right, Ma: I'm Only D.J.-ing," *New York Times*, April 30, 2006, AR1.

58. Steve Dougherty, "The Santa Claus of Christmas Songs," *Wall Street Journal*, December 17, 2010, www.wsj.com/articles/SB10001424052748704156304576003783102082942.

59. Christopher Phillips, "FROM HIS HOME TO YOURS, S01 E04: Live Every Day as If You're Going to Live Forever," *Backstreets.com*, May 20, 2020, www.backstreets.com/newsarchive107.html.

60. "Volume 1," *From My Home to Yours*, April 2020.

61. David Brooks, "Bruce Springsteen's Playlist for the Trump Era," *The Atlantic*, June 23, 2020, www.theatlantic.com/ideas/archive/2020/06/bruce-springsteens-playlist/613378.

62. It was also named XM Underground and then XM University. See Lee Abrams, "Random Stuff from the XM Halls," *Lee Abrams Media Visions*, August 21, 2006, www.leeabramsmediavisions.com/blog/random-stuff-from-the-xm-halls.

63. This news was shared by on-air host Josiah Lambert (@JosiahOnAir), "Heya hipsters. Needed to take a sec to double check the deets on this. After a very, very healthy run, Carles has indeed decided to retire the show. I was a daily HRO reader," October 29, 2022, twitter.com/JosiahOnAir/status/1586427132603162624.

64. Louis Miller, "College and Satellite Radio Team Up," *CMJ New Music Report*, August 4, 2003, 4.

65. Advertisement in *CMJ New Music Report*, October 27, 2003, 49.

66. Jennifer Waits, "KCSU Playlist from Yesterday's XMU Student Exchange," *Spinning Indie*, May 5, 2008, http://spinningindie.blogspot.com/2008/05/kcsu-playlist-from-yesterdays-xmu.html.

67. Jennifer Waits, "XMU Seeking College Radio Stations for 'Student Exchange,'" *Spinning Indie*, September 4, 2008, http://spinningindie.blogspot.com/2008/09/xmu-seeking-college-radio-stations-for.html.

68. "Christine Sanley/Music Director/KSLU/St. Louis, MO," *CMJ*, March 18, 2008, www.cmj.com/relay/?p=3773 / https://web.archive.org/web/20080326180919/http://www.cmj.com/relay/?p=3773.

69. Jennifer Waits, "Staff Cuts at XMU," *Spinning Indie*, October 17, 2008, http://spinningindie.blogspot.com/2008/10/staff-cuts-at-xmu.html.

70. Thorburn and Jenkins, introduction, 3.

71. Jonah Weiner, "The Voices: Towards a Critical Theory of Podcasting," *Slate*, December 14, 2014, www.slate.com/articles/arts/ten%5fyears%5fin%5fyour%5fears/2014/12/what%5fmakes%5fpodcasts%5fso%5faddictive%5fand%5fpleasurable.html.

72. Razlogova, "Past and Future of Music Listening," 71.

73. "Turquoise Wisdom: Floating/A Mixtape," *Aquarium Drunkard*, February 3, 2015, www.aquariumdrunkard.com/2015/02/03/turquoise-wisdom-floating-a-mixtape.

74. As of this writing, XMU has played thirteen songs more than one hundred times in the last month, all released in 2022.

75. SiriusXM Editors, "Phoebe Bridgers Shares Music & More on New SiriusXM Show 'Saddest Factory Radio,'" *SiriusXM.ca*, March 1, 2022, www.siriusxm.ca/blog/phoebe-bridgers-shares-music-more-on-new-siriusxm-show-saddest-factory-radio.

76. Michele Amabile Angermiller, "'Cousin Brucie' to Exit Sirius XM's 60's on 6 Channel after 15 Years on the Air," *Variety*, July 20, 2020, https://variety.com/2020/music/news/cousin-brucie-sirius-xm-60s-on-6-1234720590.
77. "'Ask Cousin Brucie'—WABC," *Teen Life*, 1963, 42, 44.
78. Howard Colson, *New York Journal-American*, September 23, 1961, 24.
79. Lou O'Neill Jr., "Party Ends for Cousin Brucie after 16 Radio Rockin' Years," *New York Post*, July 19, 1977.
80. Rob Patterson, supplement, *The Soho Weekly News*, August 18, 1977.
81. Patterson, supplement.
82. Gary Lycan, "'Cousin Brucie' Morrow Will Guest Host at KODJ," *The Orange County Register*, July 3, 1989, F5.
83. Lycan, "'Cousin Brucie' Morrow."
84. David Hinckley, "Brucie: 'CBS-FM Isn't Sirius about the Oldies," *Daily News*, September 27, 2007.
85. Ben Sisario, "Cousin Brucie Stays on Satellite," *New York Times*, September 26, 2007, E2.
86. David Hinckley, "Brucie: 'CBS-FM Isn't Sirius about the Oldies," *Daily News*, September 27, 2007.
87. "'Cousin Brucie' Morrow: Historic First Program on Satellite Radio!," broadcast by Sirius XM from the Rock & Roll Hall of Fame in Cleveland, OH, July 1, 2005.
88. "Deep Tracks," *SiriusXM*, www.siriusxm.ca/channels/deep-tracks.
89. Dennis McDougal, "Dusty Street," *Los Angeles Times*, October 13, 1985, www.latimes.com/archives/la-xpm-1985-10-13-tm-15520-story.html.
90. "Deep Tracks," *SiriusXM*, www.siriusxm.ca/channels/deep-tracks.
91. Andy St. Pierre, "Interview: Meg Griffin, The DJ!," *Allston Pudding*, http://allstonpudding.com/interview-meg-griffin/?pps=full_post.
92. "Howard Reunited with Meg Griffin," *HowardStern.com*, October 30, 2007, https://strn.it/OyGAEk.
93. Andy St. Pierre, "Interview: Meg Griffin, The DJ!," Allston Pudding, http://allstonpudding.com/interview-meg-griffin/?pps=full_post.
94. This isn't to say that all of their music careers have ended, but they have significantly changed and are much less tied to the production and performance of new music.
95. Matt Diehl, "Great Expectations: LL Cool J," *Rolling Stone*, September 28, 2000, 58.
96. SiriusXM, "LL COOl J Launches His Exclusive New SiriusXM Channel 'Rock the Bells Radio' on March 28," *PR Newswire*, March 27, 2018, www.prnewswire.com/news-releases/ll-cool-j-launches-his-exclusive-new-siriusxm-channel-rock-the-bells-radio-on-march-28-300620162.html.
97. According to https://xmplaylist.com in May 2024.

98. SiriusXM, "Disco and Old School Hip-Hop Channels Back by Popular Demand on SIRIUS XM Radio," December 16, 2008, https://web.archive.org/web/20081219222951/http://xmradio.mediaroom.com/index.php?s=press_releases&item=1677.

99. As described on the channel website, www.siriusxm.com/channels/rock-the-bells-radio.

100. All Access Music Group, "50s on 5 and 90s on 9 Channels Go Jock-Free on SiriusXM," *All Access*, February 3, 2014, www.allaccess.com/net-news/archive/story/126429/50s-on-5-and-90s-on-9-channels-go-jock-free-on-sir?ref=mail_recap.

101. "Lisa Loeb Joins SiriusXM's 90s on 9," *RadioInsight.com*, May 23, 2022, https://radioinsight.com/headlines/226604/lisa-loeb-joins-siriusxms-90s-on-9.

102. Caryn James, "Coming of Age in Snippets: Life as a Twentysomething," *New York Times*, February 18, 1994, C3.

103. Candice Frederick, "Lisa Loeb Paved Her Own Way in the '90s—A Rare Feat for Independent Artists of the Time," *HuffPost*, August 31, 2022, www.huffpost.com/entry/lisa-loeb-90s-music-profile_n_630d2734e4b088f74236a534.

104. Lucia Bohórquez, "Memories of Two Punk Legends, Marky Ramone and Clem Burke: 'There's Nothing Too Negative about Fame; Money Always Follows,'" *El País*, April 25, 2022, https://english.elpais.com/culture/2022-04-25/memories-of-two-punk-legends-marky-ramone-and-clem-burke-theres-not-much-negative-about-fame-money-always-follows.html.

105. According to "Marky Ramone's Punk Rock," *xmplaylist.com*, https://xmplaylist.com/station/markyramonespunkrock/most-heard.

106. Some of the other radio hosts, like the MTV VJs, are said to deliver the programs from their own home studios, a practice more common now after COVID pandemic lockdowns.

107. Tammy La Gorce, "How Marky Ramone, Punk Rocker, Spends His Sundays," *New York Times*, December 14, 2018, www.nytimes.com/2018/12/14/nyregion/how-marky-ramone-punk-rocker-spends-his-sundays.html.

108. Bob Ruggiero, "Marky Ramone Gabba Gabbas Away in New Memoir," *Houston Press*, January 22, 2015, www.houstonpress.com/music/marky-ramone-gabba-gabbas-away-in-new-memoir-6493568.

109. Michael Saponara, "LL Cool J launches Classic Hip-Hop Radio Channel 'Rock the Bells' on SiriusXM," *XXL*, March 29, 2018, www.xxlmag.com/ll-cool-rock-the-bells-siriusxm.

110. Meyers, *Word from Our Sponsor*, 201, 204.

111. SiriusXM, "Meet SiriusXM's New Recording Academy Members: Rida Naser & Ronnie Triana," *Blog.SiriusXM*, July 6, 2022, https://blog.siriusxm.com/recording-academy-members.

112. Anonymous, interview by Brian Fauteux, November 11, 2016.

113 Anonymous, interview, November 11, 2016.

114. Anonymous, interview, November 11, 2016.
115. Anonymous, interview, November 11, 2016.

6. THE TRANSNATIONAL, TECHNOLOGICAL, AND PROGRAMMING EXPANSION OF SIRIUSXM

1. SiriusXM Radio Inc., *Proxy Statement and 2009 Annual Report*, 2010.
2. "Merging Ahead," *Billboard*, November 6, 2010, 11.
3. Bottomley, *Sound Streams*, 234.
4. Hilmes, foreword, vi.
5. Leila Cobo, "Latin Notas: Everybody Get Up: Pitbull's Hitmaking Prowess Crosses Genre Lines," *Billboard*, July 17, 2010, 13.
6. Angie Romero, "Pitbull to Launch His Own SiriusXM Channel," *Billboard*, March 25, 2015, www.billboard.com/music/latin/pitbull-siriusxm-channel-globalization-6509697.
7. Angie Romero, "Pitbull to Launch Two New Original Series on Endemol Beyond Digital Network," *Billboard*, May 8, 2015, www.billboard.com/music/latin/pitbull-endemol-new-series-masterclass-pitbull-presents-6561030.
8. "FCC Okays Sirius XM Repeaters for Puerto Rico," *Radio World*, September 14, 2009, www.radioworld.com/news-and-business/fcc-okays-sirius-xm-repeaters-for-puerto-rico.
9. "Satellite Radio Service Near Mexican Border Protected," *Billboard*, August 12, 2000, 70.
10. As of this writing, debate about how to regulate content on streaming media surrounds Bill C-11 or the Online Streaming Act.
11. Broadcasting Act, S.C. 1991, https://laws.justice.gc.ca/eng/acts/b-9.01/FullText.html.
12. For musical selections, Canadian content is determined by the MAPL system. To qualify as Canadian, a musical selection must fulfill at least two of the MAPL categories, in which "M" refers to the composer of the music, "A," to the nationality of the artist, "P" to the location of the recorded performance or a live performance that is broadcast, and "L" to the nationality of the person who wrote the lyrics.
13. Canada, *Royal Commission on Radio Broadcasting*, Ottawa, 1929.
14. Armstrong, *Broadcasting Policy in Canada*, 225.
15. Armstrong, *Broadcasting Policy in Canada*, 225.
16. Dorland, introduction, ix–x.
17. Berland, "Mapping Space," 131.
18. Kirk, *Satellite Communications in Canada*, 221.
19. O'Neill and Murphy, "Crossing Borders," 184.
20 O'Neill and Murphy, "Crossing Borders," 177.

21. Galperin, "Cultural Industries."
22. Stephens, "Telecommunications under NAFTA," 95.
23. XM Satellite Radio Inc., *Form 10-K: Annual Report Pursuant to Section 13 or 15(d) of the Securities Exchange Act of 1934, for the Fiscal Year Ended December 31, 2005*, 2006, 16.
24. Sirius Satellite Radio, *2006 Annual Report and Proxy Statement*, 2007, F-20.
25. Larry LeBlanc, "Canadian Radio Mulls Satellite Regulations," *Billboard*, July 2, 2005, 13.
26. Larry LeBlanc, "CanCon to Make Waves," *Billboard*, March 4, 2006, 12.
27. Larry LeBlanc, "Canada's on the Beam," *Billboard*, February 4, 2006, 12.
28. Andrea Warner, "CBC Radio 3," *The Writerly Life*, October 28, 2011, http://thewriterlylife.blogspot.ca/2011/10/cbc-radio-3.html.
29. Allison Buchan-Terrell, "Radio 3 Canada's Most Popular Podcast—CBC Podcast Host Lawrence Introduces New Canadian Music," *The Gazette*, November 9, 2006, www.gazette.uwo.ca/article.cfm?articleID=1403&day=9&month=11&section=Arts&year=2006.
30. Buchan-Terrell, "Radio 3."
31. Andrea Warner, "CBC Radio 3," *The Writerly Life*, October 28, 2011, http://thewriterlylife.blogspot.ca/2011/10/cbc-radio-3.html.
32. Alexandra Gill, "CBC Plans to Beam Up Radio 3," *The Globe and Mail*, November 2, 2005, www.theglobeandmail.com/arts/cbc-plans-to-beam-up-radio-3/article1130231.
33. Andrea Warner, "CBC Radio 3," *The Writerly Life*, October 28, 2011, http://thewriterlylife.blogspot.ca/2011/10/cbc-radio-3.html.
34. CRTC, *Transcript of Proceedings, Various Broadcast Applications*, November 4, 2004, www.crtc.gc.ca/eng/transcripts/2004/tb1104.htm.
35. CRTC, *Transcript of Proceedings*.
36. CRTC, *Transcript of Proceedings*.
37. CRTC, *Transcript of Proceedings*.
38. CRTC, *Transcript of Proceedings*.
39. CRTC, *Transcript of Proceedings*.
40. CRTC, *Broadcasting Public Notice CRTC 2005-61*, Ottawa, 2005, www.crtc.gc.ca/eng/archive/2005/pb2005-61.htm.
41. CRTC, *Broadcasting Public Notice*.
42. CRTC, *Transcript of Proceedings, to Consider the Broadcasting Applications Listed in Broadcasting Notice of Consultation CRTC 2012-224 and 2012-224-1*, Gatineau, June 21, 2012, www.crtc.gc.ca/eng/transcripts/2012/tb0621.html.
43. CRTC, *Broadcasting Decision CRTC 2011-240*, Ottawa, 2011, www.crtc.gc.ca/eng/archive/2011/2011-240.htm.
44. CRTC, *Broadcasting Decision CRTC 2011-240*.
45. CRTC, *Broadcasting Decision CRTC 2011-240*.

NOTES 271

46. A Canadian-produced channel was defined in 2005 as "a channel produced in Canada that consists of programming not less than 50% of which is produced and broadcast for the first time on that channel." This requirement was increased to 70% by the CRTC in 2012. For more, see CRTC, *Broadcasting Decision CRTC 2012-629*, Ottawa, 2012, http://www.crtc.gc.ca/eng/archive/2012/2012-629.htm.

47. The CRTC defines a hit as "any selection that, up to and including 31 December 1980, reached one of the Top 40 positions in the charts used by the Commission to determine hits," including *Billboard* Hot 100 Singles and *Billboard* Hot Country. See CRTC, *Circular No. 445*, Ottawa, 2001, www.crtc.gc.ca/eng/archive/2001/C2001-445.htm.

48. CRTC, *Transcript of Proceedings, to Consider the Broadcasting Applications Listed in Broadcasting Notice of Consultation CRTC 2012-224 and 2012-224-1*, Gatineau, June 21, 2012, www.crtc.gc.ca/eng/transcripts/2012/tb0621.html.

49. CRTC, *Transcript of Proceedings*.

50. CRTC, *Transcript of Proceedings*.

51. CRTC, *Transcript of Proceedings*.

52. Jonathan Ratner, "Satellite Radio's Family Squabble: Sirius XM Canada and Parent in Fight Over Money Amid Takeover Bid to Take Canadian Entity Private," *Financial Post*, May 2, 2016, https://financialpost.com/investing/market-moves/satellite-radios-family-squabble-sirius-xm-canada-and-parent-in-fight-over-money-amid-takeover-bid-to-take-canadian-entity-private.

53. "SiriusXM Enters into Agreement to Back Sirius XM Canada's Going Private Transaction," *PR Newswire*, May 13, 2016, www.prnewswire.com/news-releases/siriusxm-enters-into-agreement-to-back-sirius-xm-canadas-going-private-transaction-300268261.html.

54. The Canadian Press, "Sirius XM Canada Paves Way for CBC to Sell Its Stake," *Marketing*, May 13, 2016, http://marketingmag.ca/media/sirius-xm-canada-paves-way-for-cbc-to-sell-its-stake-174494.

55. Sirius XM Holdings Inc., *Form 10-K: Annual Report Pursuant to Section 13 or 15(d) of the Securities Exchange Act of 1934 for the Fiscal Year Ended December 31, 2017*, 2018, 27.

56. CBC Audience Services, "Changes to CBC Music and ICI Musique Content on SiriusXM," *CBC Help Centre*, November 2, 2022, https://cbchelp.cbc.ca/hc/en-ca/articles/9323838282267-Changes-to-CBC-Music-and-ICI-Musique-content-on-SiriusXM.

57. Richie Assaly, "'Final Nail in the Coffin': Why SiriusXM Dropping CBC Radio 3 Is 'Potentially Catastrophic' for Canadian Artists," *Toronto Star*, October 25, 2022, www.thestar.com/entertainment/music/2022/10/25/final-nail-in-the-coffin-why-siriusxm-dropping-cbc-radio-3-is-potentially-catastrophic-for-canadian-artists.html.

58. Assaly, "'Final Nail in the Coffin.'"

59. CRTC, *Broadcasting Decision CRTC 2019-431*, Ottawa, December 19, 2019, https://crtc.gc.ca/eng/archive/2019/2019-431.htm.

60. CRTC, *Broadcasting Regulatory Policy CRTC 2011-316*, Ottawa, May 12, 2011, https://crtc.gc.ca/eng/archive/2011/2011-316.htm.

61. CRTC, *Broadcasting Decision CRTC 2019-431*, Ottawa, December 19, 2019, https://crtc.gc.ca/eng/archive/2019/2019-431.htm.

62. Richie Assaly, "'Final Nail in the Coffin': Why SiriusXM Dropping CBC Radio 3 Is 'Potentially Catastrophic' for Canadian Artists," *Toronto Star*, October 25, 2022, www.thestar.com/entertainment/music/2022/10/25/final-nail-in-the-coffin-why-siriusxm-dropping-cbc-radio-3-is-potentially-catastrophic-for-canadian-artists.html.

63. As determined by https://xmplaylist.com.

64. Shirley Halperin and Gary Trust, "Are Hits Enough to Save Radio?," *Billboard*, January 17, 2015, 49.

65. Anonymous, interview by Brian Fauteux, November 11, 2016.

66. "Merging Ahead," *Billboard*, November 6, 2010, 11.

67. Glenn Peoples, "Easy Does It," *Billboard*, June 22, 2013, 30–32.

68. Glenn Peoples, "Tune In To The Web," *Billboard*, April 27, 2013, 10.

69. SiriusXM Holdings Inc., *Proxy Statement & 2014 Annual Report*, 2015.

70. Sirius XM Holdings Inc., *Form 10-K (Annual Report): Filed 02/04/14 for the Period Ending 12/31/13*, 2014, 10.

71. Jackie Madrigal, "The State of Satellite Radio: XM and Sirius Court Hispanic Listeners," *R&R*, November 10, 2006, 66.

72. Justino Águila, "En Breve," *Billboard*, November 5, 2011, 10.

73. Leila Cobo, "En Breve," *Billboard*, August 25, 2012, 8.

74. Lelia Cobo, "Latin Notas," *Billboard*, July 20, 2013, 12–13.

75. "Further Dealings," *Billboard*, April 27, 2013, 13.

76. Glenn Peoples, "Radio's Personal Touch," *Billboard*, May 18, 2013, 18–19.

77. Sirius XM Holdings Inc., *Form 10-K (Annual Report): Filed 02/05/15 for the Period Ending 12/31/14*, 2015, 2.

78. Bull, *Sound Moves*, 2.

79. Antony Bruno, "Are Downloads in Satellite Radio's Future?," *Billboard*, July 23, 2005, 18.

80. Antony Bruno, "Digital Drive," *Billboard*, January 12, 2008, 31–32.

81. Steve Morgenstern, "12 Cool Tech Toys," *Billboard*, May 17, 2008, 26–27.

82. Antony Bruno, "Sirius Solution," *Billboard*, January 9, 2010, 10.

83. SiriusXM Satellite Radio, *Proxy Statement and 2011 Annual Report*, 2012, 22.

84. Anonymous, interview by Brian Fauteux, November 11, 2016.

85. Poell, Nieborg, and Duffy, *Platforms and Cultural Production*, 5.

86. Van Dijck, Poell, and de Waal, *Platform Society*, 40–47.

87. Bonini and Gandini, "'First Week Is Editorial,'" 2.

88. Morris, "Music Platforms and Optimization of Culture," 2.
89. Magaudda, "Smartphones, Streaming Platforms."
90. Peter Wayner, "Monitoring Your Health with Mobile Devices," *New York Times*, February 22, 2012, www.nytimes.com/2012/02/23/technology/personaltech/monitoring-your-health-with-mobile-devices.html.
91. Tudor, "Who Counts?," 839.
92. Hannam, Sheller, and Urry, "Editorial," 1.
93. See Srnicek, *Platform Capitalism*; Pedersen, "Datafication and Ubiquitous Listening"; and Gallego, "Value of Sound."
94. Anonymous, interview by Brian Fauteux, November 11, 2016.
95. SiriusXM Canada, "Privacy Policy," *SiriusXM.ca*, September 1, 2015, www.siriusxm.ca/privacy-policy.
96. "Further Dealings," *Billboard*, April 27, 2013, 13.
97. Alex Pham, "Mel Karmazin to Step Down from SiriusXM," *Billboard*, November 3, 2012, 5.

7. EMBEDDED RADIO AND THE LIMITS OF EXPANSION

1. Suisman, *Selling Sounds*, 51.
2. Frith, "Industrialization of Music," 231.
3. Alex Knapp, "Elon Musk's First Astronaut Launch Is One Giant Leap for Space Capitalism," *Forbes*, May 25, 2020, www.forbes.com/sites/alexknapp/2020/05/25/elon-musks-first-astronaut-launch-is-one-giant-leap-for-space-capitalism; and Paris Marx, "Yes to Space Exploration: No to Space Capitalism," *Jacobin*, 2020, https://jacobin.com/2020/06/spacex-elon-musk-jeff-bezos-capitalism.
4. Shammas and Holen, "One Giant Leap for Capitalistkind," 8.
5. Tim Jackson, "Billionaire Space Race: The Ultimate Symbol of Capitalism's Flawed Obsession with Growth," *The Conversation*, July 20, 2021, https://theconversation.com/billionaire-space-race-the-ultimate-symbol-of-capitalisms-flawed-obsession-with-growth-164511.
6. Alex Knapp, "Elon Musk's First Astronaut Launch Is One Giant Leap for Space Capitalism," *Forbes*, May 25, 2020, www.forbes.com/sites/alexknapp/2020/05/25/elon-musks-first-astronaut-launch-is-one-giant-leap-for-space-capitalism.
7. Paris Marx, "Yes to Space Exploration: No to Space Capitalism," *Jacobin*, 2020, https://jacobin.com/2020/06/spacex-elon-musk-jeff-bezos-capitalism.
8. Tim Jackson, "Billionaire Space Race: The Ultimate Symbol of Capitalism's Flawed Obsession with Growth," *The Conversation*, July 20, 2021, https://theconversation.com/billionaire-space-race-the-ultimate-symbol-of-capitalisms-flawed-obsession-with-growth-164511.

9. Jason Rainbow, "SiriusXM Orders Pair of Satellites to Expand in Canada and Alaska," *SpaceNews*, November 29, 2022, https://spacenews.com/siriusxm-orders-pair-of-satellites-to-expand-in-canada-and-alaska.

10. "Census Bureau Releases New Brief about Travel to Work Since Pandemic's Onset," *Census.gov*, February 20, 2024, www.census.gov/newsroom/press-releases/2024/travel-to-work-since-pandemic.html.

11. Rich Appel et al., "Digital Power Players 2017," *Billboard*, September 2, 2017, 57.

12. Anna Washenko, "Sirius XM Launches 'Essential,' an Out-of-Car Streaming Subscription Service," *RAIN News*, April 22, 2019, https://rainnews.com/sirius-xm-launches-essential-an-out-of-car-streaming-subscription-service.

13. Brad Hill, "SiriusXM Unleashes Largest-Ever Media Campaign, Emphasizing Out-of-Car Listening," *RAIN News*, September 7, 2021, https://rainnews.com/siriusxm-unleashes-largest-ever-media-campaign-emphasizing-out-of-car-listening.

14. "The Home of SiriusXM Presents: Gizmos," posted September 6, 2021, by SiriusXM, YouTube, www.youtube.com/watch?v=0cwXCEWCGSA&t=27s&ab_channel=SiriusXM.

15. Yinka Adegoke, "Risk Taker," *Billboard*, October 20, 2012, 8.

16. Adegoke, "Risk Taker," 8.

17. SiriusXM Holdings Inc., *Proxy Statement & 2013 Annual Report*, 2014, 1.

18. David Bank, "Sirius Rabid for Stock Buyback," *Variety*, March 26, 2013, 67.

19. "Further Dealings," *Billboard*, July 20, 2013, 11.

20. SiriusXM Holdings Inc., *Proxy Statement & 2013 Annual Report*, 2014.

21. "Liberty Heckled over Sirius Bid," *Billboard*, January 18, 2014, 6.

22. "SiriusXM Revenue Rockets," *Billboard*, February 15, 2014, 6.

23. SiriusXM Holdings Inc., *Proxy Statement & 2015 Annual Report*, 2016, 23; and Sirius XM Holdings Inc., *Schedule 14A: Proxy Statement Pursuant to Section 14(a) of the Securities Exchange Act of 1934*, 2021, 29.

24. DeWaard, Fauteux, and Selman, "Independent Canadian Music," 252.

25. Stahl, "Primitive Accumulation"; Hesmondhalgh, "User-Generated Content"; and Hesmondhalgh and Baker, "'Complicated Version of Freedom.'"

26. DeWaard, Fauteux, and Selman, "Independent Canadian Music," 252.

27. Burkart, "Loose Integration in Popular Music Industry."

28. Prey, Del Valle, and Zwerver, "Platform Pop," 85.

29. DeWaard, Fauteux, and Selman, "Independent Canadian Music."

30. Bottomley, *Sound Streams*, 27.

31. Glenn Peoples, "Music Subscription Battle Gets Real," *Billboard*, January 26, 2013, 7.

32. Glenn Peoples, "Leading from the Front," *Billboard*, May 4, 2013, 28–29.

33. Marc Hogan, "Beats 1: Back to the Future?," *Billboard*, August 22, 2015, 42–43.

34. Belinfante and Johnson, "Competition, Pricing and Concentration,"12, 19; and Negus, "Plugging and Programming," 57.

35. Fauteux, *Music in Range*; and Kruse, *Site and Sound*.

36. Hendy, "Pop Music Radio," 743.

37. Frank DiGiacomo, "The Power 100," *Billboard*, February 14, 2015, 80.

38. "Top Pop Programmers 2017," *Billboard*, July 14, 2017, 43–44.

39. Gary Trust, "A New Edition of Boy Bands?," *Billboard*, March 31, 2012, 27.

40. Dan Hyman, "AJR 'Ready' for Stardom," *Billboard*, September 28, 2013, 44.

41. "Dance Power Players," *Billboard*, June 24, 2017, 45.

42. Emily Zemler, "Growing 'Tally,'" *Billboard*, September 14, 2013, 44.

43. "OVER Heard," *Billboard*, October 26, 2013, 21.

44. "40 under 40," *Billboard*, October 3, 2015, 60.

45. Jon Porter, "Spotify Is First Music Streaming Service to Surpass 200M Paid Subscribers," *The Verge*, January 31, 2023, www.theverge.com/2023/1/31/23577499/spotify-q4-2022-earnings-release-subscriber-growth-layoffs.

46. Prey, Del Valle, and Zwerver, "Platform Pop," 75.

47. Bottomley, *Sound Streams*, 145–147.

48. Steve Knopper, "Radio's Debt Spiral," *Billboard*, May 28, 2016, 11.

49. Knopper, "Radio's Debt Spiral," 12.

50. "Noted," *Billboard*, June 24, 2017, 16.

51. Melinda Newman, "Top Music Lawyers," *Billboard*, July 22, 2017, 66.

52. Hannah Karp, "Pandora Tools Help Artists Sell Their Music," *Wall Street Journal, Europe*, October 18, 2016, B3.

53. Robert Levine, "From the Desk of: Roger Lynch," *Billboard*, October 7, 2017, 14–15.

54. RAIN News Staff, "Sirius XM: 'Never Lose a Listener' (Earnings Follow-Up)," *RAIN News*, April 30, 2019, https://rainnews.com/sirius-xm-never-lose-a-listener-earnings-follow-up.

55. Ed Christman, "A Serious New Contender," *Billboard*, September 29, 2018, 19.

56. Brad Hill, "SiriusXM acquires Pandora," *RAIN News*, September 24, 2018, https://rainnews.com/sirius-xm-acquires-pandora.

57. Steve Knopper, "Will Labels Have a Sirius Problem?," *Billboard*, February 9, 2019, 22.

58. Glenn Peoples, "Onboarding Pandora," *Billboard*, December 14, 2019, 29–30.

59. RAIN News Staff, "Sirius XM: 'Never Lose a Listener' (Earnings Follow-Up)," *RAIN News*, April 30, 2019, https://rainnews.com/sirius-xm-never-lose-a-listener-earnings-follow-up.

60. Brad Hill, "Sirius XM Acquires Pandora," *RAIN News*, September 24, 2018, https://rainnews.com/sirius-xm-acquires-pandora.

61. Anna Washenko, "Purge at Pandora as Sirius Plans Takeover; Roger Lynch Out," *RAIN News*, January 30, 2019, https://rainnews.com/purge-at-pandora-as-sirius-plans-takeover-roger-lynch-out.

62. Glenn Peoples, "Onboarding Pandora," *Billboard*, December 14, 2019, 29–30.

63. Brad Hill, "Sirius XM Closes Acquisition of Pandora; Jim Meyer Lays Out Strategic Bullet Points," *RAIN News*, February 1, 2019, https://rainnews.com/sirius-xm-closes-acquisition-of-pandora-jim-meyer-lays-out-strategic-bullet-points.

64. Malina Saval, "The Satellite Radio Giant Looks West for a Hollywood Presence," *Variety*, October 15, 2019, 40.

65. Saval, "Satellite Radio Giant Looks West."

66. Todd Spangler, "SiriusXM Buys Conan O'Brien's Team Coco Podcast Company for $150 Million," *Variety*, May 23, 2022, https://variety.com/2022/digital/news/siriusxm-conan-obrien-team-coco-podcast-1235275077.

67. SiriusXM Editors, "Team Coco Radio, a New Channel Produced by Conan O'Brien, to Launch on SiriusXM," *SiriusXM.ca*, November 8, 2022, www.siriusxm.ca/blog/team-coco-radio-a-new-channel-produced-by-conan-obrien-to-launch-on-siriusxm.

68. Anna Washenko, "Artist and Consumer Rights Organizations Petition Justice Department to Block Liberty Media/iHeart Deal," *RAIN News*, April 15, 2020, https://rainnews.com/artist-and-consumer-rights-organizations-petition-justice-department-to-block-liberty-media-iheart-deal.

69. Brad Hill, "Scripps Sells Stitcher to SiriusXM: Confirmed by Both Companies," *RAIN News*, July 13, 2020, https://rainnews.com/scripps-sells-stitcher-to-siriusxm-confirmed-by-both-companies.

70. Brad Hill, "SiriusXM Acquires 99% Invisible Inc.," *RAIN News*, April 26, 2021, https://rainnews.com/siriusxm-acquires-99-invisible-inc.

71. Brad Hill, "Over Five Years: Terrestrial Radio −13%, Podcasting +266% (Cumulus/Edison)," *RAIN News*, April 18, 2022, https://rainnews.com/over-five-years-terrestrial-radio-13-podcasting-266-cumulus-edison.

72. Sirius XM Holdings Inc., *Form 10-K: Annual Report Pursuant to Section 13 or 15(d) of the Securities Exchange Act of 1934 for the Fiscal Year Ended December 31, 2017*, 2018, 13.

73. Sirius XM Holdings, *Schedule 14A. Proxy Statement Pursuant to Section 14(a) of the Securities Exchange Act of 1934*, 2018.

74. Andrew Unterberger, "8: Rap Will Rise at Alternative Radio," *Billboard*, January 12, 2019, 41.

75. Frank DiGiacomo, "The Power 100," *Billboard*, February 20, 2016, 78.

76. Glenn Peoples, "Digital Domain," *Billboard*, May 14, 2011, 8.

77. Gary Graff, "A Powerful Partnership," *Billboard*, December 21, 2013, 70, 72.

78. "The Pride List," *Billboard*, August 10, 2019, 64.
79. Prey, "Locating Power in Platformization," 1.
80. Srnicek, *Platform Capitalism*, 43.
81. Poell, Nieborg, and Duffy, *Platforms and Cultural Production*, 5.
82. "Music's Digital Elite," *Billboard*, October 24, 2015, 100.
83. "40 under 40," *Billboard*, October 14, 2017, 54.
84. Thom Duffy, "Digital Power Players 2019," *Billboard*, November 16, 2019, 84.
85. Ashley Carman and Lucas Shaw, "SiriusXM to Cut Staff in Response to Faltering Growth in Revenue," *Bloomberg*, December 1, 2022, www.bloomberg.com/news/articles/2022-12-01/siriusxm-to-cut-staff-in-response-to-faltering-growth-in-revenue.
86. Andrea Zarczynski, "SiriusXM Wins Satellite Radio, Loses Pandora Subscribers—What's Next?," *Forbes*, February 5, 2020, www.forbes.com/sites/andreazarczynski/2020/02/05/siriusxm-wins-satellite-radio-loses-pandora-subscriberswhats-next.
87. Jem Aswad, "SiriusXM Revenue and Subscribers Up, Despite Down Quarter and $976 Million Pandora Charge," *Variety*, February 2, 2021, https://variety.com/2021/digital/news/siriusxm-earnings-pandora-charge-1234898206.
88. Sirius XM Holdings Inc., *Form 10-K: Annual Report Pursuant to Section 13 or 15(d) of the Securities Exchange Act of 1934 for the Fiscal Year Ended December 31, 2020*, 2021, 8.
89. Sara Fischer, "SiriusXM Eyes Live Sports Rights as Subscriber Growth Slows," *Axios*, October 18, 2022, www.axios.com/2022/10/18/siriusxm-live-sports-rights.
90. Brad Hill, "SiriusXM Combines Acquired Sales Orgs to Create SXM Media, Reaching 150M Listeners," *RAIN News*, May 10, 2021, https://rainnews.com/siriusxm-combines-acquired-sales-orgs-to-create-sxm-media-reaching-150m-listeners.
91. Brad Hill, "Crooked Media Signs with SiriusXM / SXM Media," *RAIN News*, March 25, 2022, https://rainnews.com/crooked-media-signs-with-siriusxm-sxm-media.
92. Brad Hill, "SXM Podcast Universe Offers and Interactive Podcast Browser for Advertisers, with Stats," *RAIN News*, September 15, 2021, https://rainnews.com/sxm-podcast-universe-offers-an-interactive-podcast-browser-for-advertisers-with-stats.
93. Brad Hill, "Branded Podcasts, Sure: How about Branded Podcast Episodes? An SXM Media Example," *RAIN News*, July 6, 2021, https://rainnews.com/branded-podcasts-sure-how-about-branded-podcast-episodes-an-sxm-media-example.
94. Brad Hill, "SiriusXM Intros AudioID via AdsWizz Subsidiary; Promises 'a New Era of Identity,'" *RAIN News*, January 31, 2022, https://rainnews

.com/siriusxm-intros-audioid-via-adswizz-subsidiary-promises-a-new-era-of-identity.

95. Sirius XM Holdings Inc., *Form 10-K: Annual Report Pursuant to Section 13 or 15(d) of the Securities Exchange Act of 1934 for the Fiscal Year Ended December 31, 2020*, 2021, 36.

96. Brad Hill, "Audio Subscription Revenue Nearly Two-Thirds of Total; Digital Audio Ad Revenue Set to Gain," *RAIN News*, December 16, 2021, https://rainnews.com/audio-subscription-revenue-nearly-two-thirds-of-total-digital-audio-ad-revenue-set-to-gain.

97. Sirius XM Holdings Inc., *Schedule 14A: Proxy Statement Pursuant to Section 14(a) of the Securities Exchange Act of 1934*, 2021.

98. Sirius XM Holdings Inc., *Schedule 14A: Proxy Statement Pursuant to Section 14(a) of the Securities Exchange Act of 1934*, 2017.

99. Nilay Patel, "SiriusXM's 360 Strategy, with CEO Jennifer Witz," *The Verge*, June 6, 2023, www.theverge.com/23750277/siriusxm-strategy-ceo-jennifer-witz-howard-stern-conan-obrien-radio-podcast-cars.

100. Star, "Ethnography of Infrastructure," 381.

101. Howard, "Embedded Media," 2.

102. Howard, "Embedded Media," 24.

103. Scannell, *Radio, Television and Modern Life*, 148, 152–153.

104. Blaakilde, "Becoming of Radio Bodies," 301.

105. Bonini and Gandini, "First Week Is Editorial."

106. As said by Twitter user Jimmie Quick (@jimmieaquick), "Stitcher app = absolute best and only way I listen to pods.... you aren't providing any details about how to get my pods over into that platform," June 28, 2023, https://twitter.com/jimmieaquick/status/1674029835398840320.

107. Russ Crupnick, "Russ Crupnick: Does SiriusXM Deserve More Respect?," *RAIN News*, April 27, 2023, https://rainnews.com/russ-crupnick-siriusxm-04272023.

108. "New Users Can Get a Personalized Music Experience from the First Time They Open the SXM App with 'Channels You Might Like,'" *SiriusXM*, February 14, 2023, https://blog.siriusxm.com/sxm-app-channels-you-might-like.

109. Chris Huber, "The Meaning of Wilco's 'Jesus, Etc.,'" *Extra Chill*, December 15, 2021, https://extrachill.com/2021/12/wilco-jesus-etc-meaning.html.

110. "100 Best Songs of the 2000s," *Rolling Stone*, June 17, 2011, www.rollingstone.com/music/music-lists/100-best-songs-of-the-2000s-153056.

Bibliography

Acland, Charles R. *American Blockbuster: Movies, Technology, and Wonder*. Durham, NC: Duke University Press, 2020.

———. "Introduction: Residual Media." In *Residual Media*, edited by Charles R. Acland, xiii–xxvii. Minneapolis: University of Minnesota Press, 2007

Adorno, Theodor. *The Culture Industry: Selected Essays on Mass Culture*. New York: Routledge, 1991. First published 1938.

———. "On Popular Music." In *On Record: Rock, Pop, and the Written Word*, edited by Simon Frith and Andrew Goodwin, 301–314. New York: Pantheon Books, 1990.

Alper, Garth. "XM Reinvents Radio's Future." *Popular Music and Society* 29, no. 5 (2006): 505–518.

Anderson, Benedict. *Imagined Communities: Reflections on the Origins and Spread of Nationalism*. New York: Verso, 1991.

Andrejevic, Mark. *Infoglut: How Too Much Information Is Changing the Way We Think and Know*. New York: Routledge, 2013.

Arditi, David. "Digital Subscriptions: The Unending Consumption of Music in the Digital Era." *Popular Music and Society* 41, no. 3 (2018): 302–318.

Arendt, Hannah. *The Human Condition*. 2nd ed. Chicago: University of Chicago Press, 1998. First published 1958.

Armstrong, Robert. *Broadcasting Policy in Canada*. Toronto: University of Toronto Press, 2010.

Attali, Jacques. *Noise: The Political Economy of Music*. Translated by Brian Massumi. Minneapolis: University of Minnesota Press, 2009. First published 1985.

Auslander, Philip. *Liveness: Performance in a Mediatized Culture*. New York: Routledge, 1999.

Baade, Christina L. *Victory through Harmony: The BBC and Popular Music in World War II*. New York: Oxford University Press, 2012.

Baker, Sarah, and Alison Huber. "Saving 'Rubbish': Preserving Popular Music's Material Culture in Amateur Archives and Museums." In *Sites of Popular Music Heritage: Memories, Histories, Places*, edited by Sara Cohen, Robert Knifton, Marion Leonard, and Les Roberts, 112–124. New York: Routledge, 2014.

Barnett, Kyle. *Record Cultures: The Transformation of the U.S. Recording Industry*. Ann Arbor: University of Michigan Press, 2020.

Baudrillard, Jean. *Simulations*. Translated by Paul Foss, Paul Patton, and Philip Beitchman. New York: Semiotext[e], 1983.

———. *The System of Objects*. Translated by James Benedict. New York: Verso, 1996. First published 1968.

Beaster-Jones, Jayson. *Music Commodities, Markets, and Values: Music as Merchandise*. New York: Routledge, 2016.

Belinfante, Alexander, and Richard L. Johnson. "Competition, Pricing and Concentration in the U.S. Recorded Music Industry." *Journal of Cultural Economics* 6, no. 2 (1982): 11–24.

Bennett, Andy. "'Heritage Rock': Rock Music, Representation and Heritage Discourse." *Poetics* 37 (2009): 474–489.

Berland, Jody. "Locating Listening: Technological Space, Popular Music, and Canadian Mediations." In *The Place of Music*, edited by Andrew Leyshon, David Matless, and George Revill, 129–150. New York: Guilford Press, 1998.

———. "Mapping Space: Imaging Technologies and the Planetary Body." In *Technoscience and Cyberculture*, edited by Stanley Aronowitz, Barbara Martinsons, and Michael Menser, 123–137. New York: Routledge, 1996.

———. *North of Empire: Essays on the Cultural Technologies of Space*. Durham, NC: Duke University Press, 2009.

———. "Radio Space and Industrial Time: Music Formats, Local Narratives and Technological Mediation." *Popular Music* 9, no. 2 (1990): 179–192.

Berman, Stuart. *This Book Is Broken: The Broken Social Scene Story*. Toronto: Anansi, 2009.

Bijsterveld, Karin, and Annelies Jacobs. Introduction to *Sound Souvenirs: Audio Technologies, Memory and Cultural Practice*, edited by Karin Bijsterveld and José van Dijck, 11–21. Amsterdam: Amsterdam University Press, 2009.

Bijsterveld, Karin, Eefje Cleophas, Stefan Krebs, and Gijs Mom. *Sound and Safe: A History of Listening behind the Wheel*. New York: Oxford University Press, 2014.

Blaakilde, Anna Leonora. "The Becoming of Radio Bodies." *European Journal of Cultural Studies* 21, no. 3 (2017): 290–304.
Bolter, Jay David, and Richard Grusin. *Remediation: Understanding New Media*. Cambridge, MA: MIT Press, 2000.
Bonini, Tiziano, and Alessandro Gandini. "'First Week Is Editorial, Second Week Is Algorithmic': Platform Gatekeepers and the Platformization of Music Curation." *Social Media + Society* (October 2019): 1–11.
Bottomley, Andrew. *Sound Streams: A Cultural History of Radio-Internet Convergence*. Ann Arbor: University of Michigan Press, 2020.
Breihan, Tom. *The Number Ones: Twenty Chart-Topping Hits that Reveal the History of Pop Music*. New York: Hachette Books, 2022.
Brunet, François. "Introduction: No Representation without Circulation." In *Circulation*, edited by François Brunet, 10–39. Chicago: Terra Foundation for American Art, 2017.
Bull, Michael. "The Auditory Nostalgia of iPod Culture." In *Sound Souvenirs: Audio Technologies, Memory and Cultural Practice*, edited by Karin Bijsterveld and José van Dijck, 83–93. Amsterdam: Amsterdam University Press, 2009.
———. "Automobility and the Power of Sound." *Theory, Culture & Society* 21, nos. 4–5 (2004): 243–259.
———. *Sound Moves: iPod Culture and Urban Experience*. New York: Routledge, 2007.
Burkart, Patrick. "Loose Integration in the Popular Music Industry." *Popular Music* 28, no. 4 (2005): 489–500.
Bürkner, Hans-Joachim, and Bastian Lange. "Sonic Capital and Independent Urban Music Production: Analysing Value Creation and 'Trial and Error' in the Digital Age." *City, Culture and Society* 10 (2017): 33–40.
Caldwell, John T. "Second Shift Media Aesthetics: Programming, Interactivity, and User Flows." In *New Media: Theories and Practices of Digitextuality*, edited by Anna Everett and John T. Caldwell, 127–144. New York: Routledge, 2003.
Camille, Michael. "Simulacrum." In *Critical Terms for Art History*, edited by Robert S. Nelson and Richard Shiff, 31–44. Chicago: University of Chicago Press, 1996.
Child, Ben. "Raised in the Country, Working in the Town: Temporal and Spatial Modernisms in Bob Dylan's '*Love and Theft*.'" *Popular Music and Society* 32, no. 2 (2009): 199–210.
Chow, Rey. *The Age of the World Target: Self-Referentiality in War, Theory, and Comparative Work*. Durham, NC: Duke University Press, 2006.
Clarke, Arthur C. Preface to *The Beginnings of Satellite Communication*, by John Robinson Pierce, v–vii. San Francisco: San Francisco Press, 1968.
Clover, Joshua. *Roadrunner*. Durham, NC: Duke University Press, 2021.
Coddington, Amy. *How Hip Hop Became Hit Pop: Radio, Rap and Race*. Oakland: University of California Press, 2023.

Collis, Christy. "The Geostationary Orbit: A Critical Legal Geography of Space's Most Valuable Real Estate." *Sociological Review* 57, supp. 1 (2009): 47–65.

Corbett, John. *Extended Play: Sounding Off from John Cage to Dr. Funkenstein*. Durham, NC: Duke University Press, 1994.

Cwynar, Christopher. "Brick, Mortar, and Screen: Networked Digital Media, Popular Music, and the Reinvention of the Public Radio Station." *Journal of Radio & Audio Media* 27, no. 1 (2020): 74–92.

Dahlman, Ian. "'A Big Beautiful Mess': Collectivity, Capitalism, Arts & Crafts and Broken Social Scene." Master's thesis, Ryerson and York University, 2009.

Deleuze, Gilles. "Plato and the Simulacrum." Translated by Rosalind Krauss. *October* 27 (1983): 45–56.

Denisoff, R. Serge. *Tarnished Gold: The Record Industry Revisited*. New Brunswick, NJ: Transaction Books, 1986.

DeNora, Tia. *Music in Everyday Life*. New York: Cambridge University Press, 2000.

deWaard, Andrew. *Derivative Media: How Wall Street Devours Culture*. Oakland: University of California Press, 2024.

deWaard, Andrew, Brian Fauteux, and Brianne Selman. "Independent Canadian Music in the Streaming Age: The Sound from above (Critical Political Economy) and below (Ethnography of Musicians)." *Popular Music and Society* 45, no. 3 (2022): 251–278.

Dorland, Michael. Introduction to *The Cultural Industries in Canada: Problems, Policies and Prospects*, edited by Michael Dorland, ix–xii. Toronto: Lorimer, 1996.

Douglas, Susan J. *Inventing American Broadcasting, 1899–1922*. Baltimore, MD: Johns Hopkins University Press, 1989.

———. *Listening In: Radio and the American Imagination*. Minneapolis: University of Minnesota Press, 2004.

Drott, Eric. *Streaming Music, Streaming Capital*. Durham, NC: Duke University Press, 2024.

Dylan, Bob. *Chronicles: Volume One*. Toronto: Simon & Schuster, 2004.

Edgerton, David. *The Shock of the Old: Technology and Global History since 1900*. New York: Oxford University Press, 2007.

Editorial Collective. "Welcome to Media Industries." *Media Industries Journal* 1, no. 1 (2014): 1–5.

Elliott, Richard. "The Same Distant Places: Bob Dylan's Poetics of Place and Displacement." *Popular Music and Society* 32, no. 2 (2009): 249–270.

Eriksson, Maria, Rasmus Fleischer, Anna Johansson, Pelle Snickars, and Patrick Vonderau. *Spotify Teardown: Inside the Black Box of Streaming Media*. Cambridge, MA: MIT Press, 2019.

Fauteux, Brian. *Music in Range: The Culture of Canadian Campus Radio*. Waterloo, ON: Wilfrid Laurier University Press, 2015.

Featherstone, Mike. "Automobilities: An Introduction." *Theory, Culture & Society* 21, nos. 4–5 (2004): 1–24.
Fisher, Mark. *K-Punk: The Collected and Unpublished Writings of Mark Fisher (2004–2016)*. Edited by Darren Ambrose. London: Repeater Books, 2018.
Frith, Simon. "The Industrialization of Music." In *The Popular Music Studies Reader*, edited by Andy Bennett, Barry Shank, and Jason Toynbee, 231–238. New York: Routledge, 2000.
———. "Look! Hear! The Uneasy Relationship of Music and Television." *Popular Music* 21, no. 3 (2002): 277–290.
———. *Performing Rites: On the Value of Popular Music*. Cambridge, MA: Harvard University Press, 1996.
Gallego, J. Ignacio. "The Value of Sound: Datafication of the Sound Industries in the Age of Surveillance and Platform Capitalism." *First Monday* 26, no. 7 (2021).
Galperin, Herman. "Cultural Industries in the Age of Free-Trade Agreements." *Canadian Journal of Communication* 24, no. 1 (1999): 49–77.
Gibson, Chris. "Recording Studios: Relational Spaces of Creativity in the City." *Built Environment* 31, no. 3 (2005): 192–207.
Gibson, Chris, and John Connell. "Music, Tourism and the Transformation of Memphis." *Tourism Geographies: An International Journal of Tourism Space, Place and Environment* 9, no. 2 (2007): 160–190.
Gitelman, Lisa. *Always Already New: Media, History and the Data of Culture*. Cambridge, MA: MIT Press, 2006.
Graham, Stephen. *Vertical: The City from Satellites to Bunkers*. New York: Verso, 2016.
Hamilton, James. "Unearthing Broadcasting in the Anglophone World." In *Residual Media*, edited by Charles R. Acland, 283–300. Minneapolis: University of Minnesota Press, 2007.
Hannam, Kevin, Mimi Sheller, and John Urry. "Editorial: Mobilities, Immobilities and Moorings." *Mobilities* 1, no. 1 (2006): 1–22.
Harvey, David. *Spaces of Global Capitalism: Towards a Theory of Uneven Geographical Development*. New York: Verso, 2006.
Hendy, David. "Pop Music Radio in the Public Service: BBC Radio 1 and New Music in the 1990s." *Media, Culture & Society* 22, no. 6 (2000): 743–761.
Hesmondhalgh, David. "User-Generated Content, Free Labour and the Cultural Industries." *ephemera* 10, nos. 3–4 (2010): 267–284.
———. *Why Music Matters*. Malden, MA: Wiley Blackwell, 2013.
Hesmondhalgh, David, and Sarah Baker. "'A Very Complicated Version of Freedom': Conditions and Experiences of Creative Labour in Three Cultural Industries." *Poetics* 38, no. 1 (2010): 4–20.
Hilmes, Michele. "Foreword: Transnational Radio in the Global Age." *Journal of Radio Studies* 11, no. 1 (2004): iii–vi.

———. "Radio and the Imagined Community." In *The Sound Studies Reader*, edited by Jonathan Sterne, 351–362. New York: Routledge, 2012.

Holbrook, Morris B., and Robert M. Schindler. "Some Exploratory Findings on the Development of Musical Tastes." *Journal of Consumer Research* 16, no. 1 (1989): 119–124.

Holt, Jennifer. *Empires of Entertainment: Media Industries and the Politics of Deregulation, 1980–1996*. New Brunswick, NJ: Rutgers University Press, 2011.

Holt, Jennifer, and Alisa Perren. "Introduction: Does the World Really Need One More Field of Study?" In *Media Industries: History, Theory, and Method*, edited by Jennifer Holt and Alisa Perren, 1–16. Malden, MA: Blackwell, 2009.

Howard, Philip N. "Embedded Media: Who We Know, What We Know, and Society Online." In *Society Online: The Internet in Context*, edited by Philip N. Howard and Steve Jones, 1–27. Thousand Oaks, CA: Sage, 2004.

Hunt, Lynn. *Measuring Time, Making History*. New York: Central European University Press, 2008.

Hutcheon, Linda. *The Politics of Postmodernism*. New York: Routledge, 1989.

Innis, Harold A. *Empire and Communications*. Toronto: Dundurn Press, 2007. First published 1950.

Jaffe, Leonard, ed. *Satellite Communications in the Next Decade: Proceedings of the 14th Goddard Memorial Symposium*. Science and Technology, vol. 44. San Diego: American Astronautical Society, 1977.

Jameson, Fredric. "Postmodernism, or, The Cultural Logic of Late Capitalism." *New Left Review* 146 (1984): 53–92.

———. *Postmodernism, or, The Cultural Logic of Late Capitalism*. Durham, NC: Duke University Press, 1991.

Jarett, David, ed. *Satellite Communications: Future Systems*. New York: American Institute of Aeronautics and Astronautics, 1977.

Jenkins, Henry, Sam Ford, and Joshua Green. *Spreadable Media: Creating Value and Meaning in a Networked Culture*. New York: New York University Press, 2013.

Jewell, Katherine Rye. *Live from the Underground: A History of College Radio*. Chapel Hill: University of North Carolina Press, 2023.

Johnson, Aaron J. "A Date with the Duke: Ellington on Radio." *Musical Quarterly* 96, nos. 3–4 (2013): 369–405.

Jones, Mike. "The Music Industry as Workplace: An Approach to Analysis." In *Cultural Work: Understanding the Cultural Industries*, edited by Andrew Beck, 147–156. New York: Routledge, 2003.

Kaplan, Caren. "Mobility and War: The Cosmic View of US 'Air Power.'" *Environment and Planning A* 38, no. 2 (2006): 395–407.

Keightley, Keir. "Reconsidering Rock." In *The Cambridge Companion to Pop and Rock*, edited by Simon Frith, Will Straw, and John Street, 109–142. Cambridge: Cambridge University Press, 2001.

———. "'Turn It Down!' She Shrieked: Gender, Domestic Space, and High Fidelity, 1948–59." *Popular Music* 15, no. 2 (1996): 149–177.

Kidd, Allison. "The Beginning of the End of the Internet Radio Royalty Dispute." *Journal of Internet Law* 9, no. 4 (2005): 15–25.

Kirk, Barrie C. *Satellite Communications in Canada*. Gloucester: Telesat Enterprises, 1993.

Klaess, John. *Breaks in the Air: The Birth of Rap Radio in New York City*. Durham, NC: Duke University Press, 2022.

Klein, Bethany. "'The New Radio': Music Licensing as a Response to Industry Woe." *Media, Culture & Society* 30, no. 4 (2008): 463–478.

Klein, Bethany, Leslie M. Meier, and Devon Powers. "Selling Out: Musicians, Autonomy, and Compromise in the Digital Age." *Popular Music and Society* 40, no. 2 (2017): 222–238.

Kornbluh, Anna. *Immediacy, or the Style of Too Late Capitalism*. New York: Verso, 2023.

Kotarba, Joseph A. "Rock 'n' Roll Music as a Timepiece." *Symbolic Interaction* 25, no. 3 (2002): 397–404.

Kramer, Michael J. *The Republic of Rock: Music and Citizenship in the Sixties Counterculture*. New York: Oxford University Press, 2013.

Kronenberg, Sam. "Royalty Rates and Exclusive Releases Threaten Music Streaming." *Southern California Interdisciplinary Law Journal* 27, no. 3 (2018): 633–656.

Kruse, Holly. *Site and Sound: Understanding Independent Music Scenes*. New York: Peter Lang, 2003.

Lacey, Kate. "Up in the Air? The Matter of Radio Studies." *Radio Journal: International Studies in Broadcast & Audio Media* 16, no. 2 (2018): 109–126.

Livingston, Tamara E. "Music Revivals: Towards a General Theory." *Ethnomusicology* 43, no. 1 (1999): 66–85.

Magaudda, Paolo. "Smartphones, Streaming Platforms, and the Infrastructuring of Digital Music Practices." In *Rethinking Music through Science and Technology Studies*, edited by Antoine Hennion and Christophe Levaux, 241–255. New York: Routledge, 2021.

Manovich, Lev. *The Language of New Media*. Cambridge, MA: MIT Press, 2001.

Marshall, Lee. *Bob Dylan: The Never Ending Star*. Malden, MA: Polity, 2007.

McCourt, Tom. "Collecting Music in the Digital Realm." *Popular Music and Society* 28, no. 2 (2005): 249–252.

McLeod, Ken. "Space Oddities: Aliens, Futurism and Meaning in Popular Music." *Popular Music* 22, no. 3 (2003): 337–355.

Meier, Leslie M. *Popular Music as Promotion: Music and Branding in the Digital Age*. Malden, MA: Polity, 2017.

Meyers, Cynthia B. *A Word from Our Sponsor: Admen, Advertising, and the Golden Age of Radio*. New York: Fordham University Press, 2013.

Michelsen, Morten, Mads Krogh, Iben Have, and Steen Kaargaard Nielsen, eds. *Tunes for All? Music on Danish Radio.* Aarhus: Aarhus University Press, 2018.

Michelsen, Morten, Mads Krogh, Steen Kaargaard Nielsen, and Iben Have, eds. *Music Radio: Building Communities, Mediating Genres.* New York: Bloomsbury Academic, 2019.

Montell, Amanda. *Cultish: The Language of Fanaticism.* New York: Harper Wave, 2021.

Moore, Allan. "Authenticity as Authentication." *Popular Music* 21, no. 2 (2002): 209–223.

Morris, Jeremy Wade. "Music Platforms and the Optimization of Culture." *Social Media + Society* 6, no. 3 (2020).

Morris, Jeremy Wade, and Devon Powers. "Control, Curation and Musical Experience in Streaming Music Services." *Creative Industries Journal* 8, no. 2 (2015): 106–122.

Moscote Freire, Ariana. "Remediating Radio: Audio Streaming, Music Recommendation and the Discourse of Radioness." *Radio Journal: International Studies in Broadcast and Audio Media* 5, nos. 2–3 (2007): 97–112.

Mullen, Megan. *The Rise of Cable Programming in the United States: Revolution or Evolution?* Austin: University of Texas Press, 2003.

Murdock, Graham. "Back to Work: Cultural Labor in Altered Times." In *Cultural Work: Understanding the Cultural Industries*, edited by Andrew Beck, 15–36. New York: Routledge, 2003.

Negus, Keith. *Bob Dylan.* Bloomington: Indiana University Press, 2008.

———. "Plugging and Programming: Pop Radio and Record Promotion in Britain and the United States." *Popular Music* 12, no. 1 (1993): 57–68.

O'Malley, Kathleen M. "Freeform Radio." In *Historical Dictionary of American Radio*, edited by Donald G. Godfrey and Frederic A. Leigh, 172–173. Westport, CT: Greenwood Press, 1998.

O'Neill, Brian, and Michael Murphy. "Crossing Borders: The Introduction and Legislation of Satellite Radio in Canada." In *Down to Earth: Satellite Technologies, Industries, and Cultures*, edited by Lisa Parks and James Schwoch, 177–193. New Brunswick, NJ: Rutgers University Press, 2012.

Ozzi, Dan. *Sellout: The Major-Label Feeding Frenzy that Swept Punk, Emo, and Hardcore (1994–2007).* New York: Mariner Books, 2021.

Parks, Lisa. *Cultures in Orbit: Satellites and the Televisual.* Durham, NC: Duke University Press, 2005.

———. "Satellites, Oil, and Footprints: Eutelsat, Kazsat, and Post-Communist Territories in Central Asia." In *Down to Earth: Satellite Technologies, Industries, and Cultures*, edited by Lisa Parks and James Schwoch, 122–140. New Brunswick, NJ: Rutgers University Press, 2012.

Parks, Lisa, and James Schwoch. Introduction to *Down to Earth: Satellite Technologies, Industries, and Cultures*, edited by Lisa Parks and James Schwoch, 1–16. New Brunswick, NJ: Rutgers University Press, 2012.

Parks, Lisa, and Nicole Starosielski. Introduction to *Signal Traffic: Critical Studies of Media Infrastructures*, 1–27. Urbana: University of Illinois Press, 2015.

Pedersen, Rasmus Rex. "Datafication and the Push for Ubiquitous Listening in Music Streaming." *Journal of Media & Communication Research* 36, no. 69 (2020): 71–89.

Peters, John Durham. *The Marvelous Clouds: Toward a Philosophy of Elemental Media*. Chicago: University of Chicago Press, 2015.

———. *Speaking into the Air: A History of the Idea of Communication*. Chicago: University of Chicago Press, 1999.

Peterson, Richard A., and Roger M. Kern. "Changing Highbrow Taste: From Snob to Omnivore." *American Sociological Review* 61, no. 5 (1996): 900–907.

Pierce, John Robinson. *The Beginnings of Satellite Communication*. San Francisco: San Francisco Press, 1968.

Plasketes, George. "Romancing the Record: The Vinyl De-Evolution and Subcultural Evolution." *Journal of Popular Culture* 26, no. 1 (1992): 109–122.

Poell, Thomas, David Nieborg, and Brooke Erin Duffy. *Platforms and Cultural Production*. Medford, MA: Polity, 2022.

Prey, Robert. "Locating Power in Platformization: Music Streaming Playlists and Curatorial Power." *Social Media + Society* 6, no. 2 (2020): 1–11.

Prey, Robert, Marc Esteve Del Valle, and Leslie Zwerwer. "Platform Pop: Disentangling Spotify's Intermediary Role in the Music Industry." *Information, Communication & Society* 25, no. 1 (2022): 74–92.

Priestman, Chris. "Narrowcasting and the Dream of Radio's Great Global Conversation." *Radio Journal: International Studies in Broadcast and Audio Media* 2, no. 2 (2004): 77–88.

Prindle, Gregory M. "No Competition: How Radio Consolidation Has Diminished Diversity and Sacrificed Localism." *Fordham Intellectual Property, Media & Entertainment Law Journal* 14, no. 1 (2003): 279–326.

Razlogova, Elena. "Freeform Radio and the History of Music Streaming." In *The Oxford Handbook of Radio Studies*, edited by Michele Hilmes and Andrew Bottomley, 22–40. New York: Oxford University Press, 2024.

———. "The Past and Future of Music Listening: Between Freeform DJs and Recommendation Algorithms." In *Radio's New Wave: Global Sound in the Digital Era*, 62–76. New York: Routledge, 2013.

Rees, David W. E. *Satellite Communications: The First Quarter Century of Service*. New York: Wiley, 1990.

Regev, Motti. "Popular Music Studies: The Issue of Musical Value." *Journal of Popular Music Studies* 4, no. 2 (1992): 22–27.

Reynolds, Simon. *Retromania: Pop Culture's Addiction to Its Own Past*. New York: Faber & Faber, 2011.

Rodman, Ron. "Radio Formats in the United States: A (Hyper)Fragmentation of the Imagination." In *Music and the Broadcast Experience: Performance,*

Production, and Audiences, edited by Christina L. Baade and James Deaville, 235–257. New York: Oxford University Press, 2016.

Roessner, Jeffrey. "Radio in Transit: Satellite Technology, Cars, and the Evolution of Musical Genres." In *21st Century Perspectives on Music, Technology, and Culture*, edited by Richard Purcell and Richard Randall, 55–71. New York: Palgrave Macmillan, 2016.

Rushkoff, Douglas. *Get Back in the Box: Innovation from the Inside Out*. New York: Collins, 2005.

Russo, Alexander. *Points on the Dial: Golden Age Radio beyond the Networks*. Durham, NC: Duke University Press, 2010.

———. "Radio Formats: Sound Rules for Addressing the Narrowcast Audience Commodity." In *The Routledge Companion to Radio and Podcast Studies*, edited by Mia Lindgren and Jason Loviglio, 198–207. New York: Routledge, 2022.

Russo, Alexander, and Bill Kirkpatrick. "Beyond the Terrestrial? Networked Distribution, Multimodal Media, and the Place of the Local in Satellite Radio." In *Down to Earth: Satellite Technologies, Industries, and Cultures*, edited by Lisa Parks and James Schwoch, 156–176. New Brunswick, NJ: Rutgers University Press, 2012.

Scannell, Paddy. *Radio, Television and Modern Life: A Phenomenological Approach*. Cambridge, UK: Blackwell, 1996.

Scannell, Paddy, and David Cardiff. *A Social History of British Broadcasting*. Vol. 1, *1922–1939: Serving the Nation*. Oxford: Basil Blackwell, 1991.

Scobie, Stephen. "Plagiarism, Bob, Jean-Luc and Me." In *Refractions of Bob Dylan: Cultural Appropriations of an American Icon*, edited by Eugen Banauch, 188–204. Manchester: Manchester University Press, 2015.

Shammas, Victor L., and Tomas B. Holen. "One Giant Leap for Capitalistkind: Private Enterprise in Outer Space." *Palgrave Communications* 5, no. 1 (2019): 1–9.

Sheller, Mimi, and John Urry. "The City and the Car." *International Journal of Urban and Regional Research* 24, no. 4 (2000): 737–757.

———. "The New Mobilities Paradigm." *Environment and Planning A* 38 (2006): 207–226.

Shields, Rob. "The Bridge Spanning Past, Present, and Future: Time Infrastructure." *Canadian Journal of Communication* 46, no. 2 (2021): 345–361.

Shumway, David R. "Rock 'n' Roll Sound Tracks and the Production of Nostalgia." *Cinema Journal* 38, no. 2 (1999): 36–51.

Skeggs, Beverley. *Class, Self, Culture*. London: Routledge, 2004.

Smith-Shomade, Beretta E. "Narrowcasting in the New World Information Order: A Space for the Audience." *Television and New Media* 5, no. 1 (2004): 69–81.

Spilker, Hendrik Storstein, and Terje Colbjørnsen. "The Dimensions of Streaming: Toward a Typology of an Evolving Concept." *Media, Culture & Society* 42, nos. 7–8 (2020): 1210–1225.

Srnicek, Nick. *Platform Capitalism*. Malden, MA: Polity, 2017.

Stahl, Matt. "Primitive Accumulation, the Social Common, and the Contractual Lockdown of Recording Artists at the Threshold of Digitalization." *Ephemera: Theory & Politics in Organization* 10, nos. 3–4 (2010): 337–356.

Star, Susan Leigh. "The Ethnography of Infrastructure." *American Behavioral Scientist* 43, no. 3 (1999): 377–391.

Stephens, Geneva E. "Telecommunications under the NAFTA and Its Effect on Canada's Telecommunications Industry." *Law and Business Review of the Americas* 3, no. 1 (1997): 93–111.

Sterne, Jonathan. "Bourdieu, Technique and Technology." *Cultural Studies* 17, nos. 3–4 (2003): 367–389.

Stiernstedt, Fredrik. "The Political Economy of the Radio Personality." *Journal of Radio & Audio Media* 21, no. 2 (2014): 290–306.

Stone, Alison. *The Value of Popular Music: An Approach from Post-Kantian Aesthetics*. Cham, Switzerland: Palgrave Macmillan, 2016.

Straw, Will. "The Circulatory Turn." In *The Wireless Spectrum: The Politics, Practices, and Poetics of Mobile Media*, edited by Barbara Crow, Michael Longford, and Kim Sawchuk, 17–28. Toronto: University of Toronto Press, 2010.

Streeter, Thomas. "Blue Skies and Strange Bedfellows: The Discourse of Cable Television." In *The Revolution Wasn't Televised: Sixties Television and Social Conflict*, edited by Lynn Spigel and Michael Curtin, 221–242. New York: Routledge, 1997.

Styvén, Maria. "The Intangibility of Music in the Internet Age." *Popular Music and Society* 30, no. 1 (2007): 53–74.

Suisman, David. *Selling Sounds: The Commercial Revolution in American Music*. Cambridge, MA: Harvard University Press, 2009.

Svec, Henry Adam. *American Folk Media as Tactical Media*. Amsterdam: Amsterdam University Press, 2018.

Tacchi, Jo. "Nostalgia and Radio Sound." In *The Auditory Culture Reader*, edited by Michael Bull and Les Back, 281–295. New York: Berg, 2003.

Taylor, Steve. "'I Am What I Play': The Radio DJ as Cultural Arbiter and Negotiator." In *Cultural Work: Understanding the Cultural Industries*, edited by Andrew Beck, 73–100. New York: Routledge, 2003.

Taylor, Timothy D. "Circulation, Value, Exchange, and Music." *Ethnomusicology* 64, no. 2 (2020): 254–273.

———. *Making Value: Music, Capital, and the Social*. Durham, NC: Duke University Press, 2024.

———. "Music and the Rise of Radio in 1920s America: Technological Imperialism, Socialization, and the Transformation of Intimacy." *Historical Journal of Film, Radio and Television* 22, no. 4 (2002): 425–443.

———. "Performance and Nostalgia on the Oldies Circuit." In *Sound Souvenirs: Audio Technologies, Memory and Cultural Practice*, edited by Karin Bijsterveld and José van Dijck, 94–106. Amsterdam: Amsterdam University Press, 2009.

———. *Strange Sounds: Music, Technology and Culture*. New York: Routledge, 2001.

Thorburn, David, and Henry Jenkins. Introduction to *Rethinking Media Change: The Aesthetics of Transition*, edited by David Thorburn and Henry Jenkins, 1–16. Cambridge, MA: MIT Press, 2003.

Toynbee, Jason. "Fingers to the Bone or Spaced out on Creativity? Labor Process and Ideology in the Production of Pop." In *Cultural Work: Understanding the Cultural Industries*, edited by Andrew Beck, 39–55. New York: Routledge, 2003.

Tschmuck, Peter. *Creativity and Innovation in the Music Industry*. New York: Springer, 2012.

Tsing, Anna. "The Global Situation." *Cultural Anthropology* 15, no. 3 (2000): 327–360.

Tudor, Deborah. "Who Counts? Who Is Being Counted? How Audience Measurement Embeds Neoliberalism into Urban Space." *Media, Culture & Society* 31, no. 5 (2009): 833–840.

van der Hoven, Arno. "Remembering the Popular Music of the 1990s: Dance Music and the Cultural Meanings of Decade-Based Nostalgia." *International Journal of Heritage Studies* 20, no. 3 (2014): 316–330.

———. "Songs that Resonate: The Uses of Popular Music Nostalgia." In *The Routledge Companion to Popular Music History and Heritage*, edited by Sarah Baker, Catherine Strong, Lauren Istvandity, and Zelmarie Cantillon, 238–246. New York: Routledge, 2018.

van Dijck, José, Thomas Poell, and Martijn de Waal. *The Platform Society*. New York: Oxford University Press, 2018.

Veal, Michael. "Starship Africa." In *The Sound Studies Reader*, edited by Jonathan Sterne, 454–467. New York: Routledge, 2012.

Watson, Jada. "Gender on the *Billboard* Hot Country Songs Chart, 1996–2016." *Popular Music and Society* 42, no. 5 (2019): 538–560.

Webster, Jack. "Taste in the Platform Age: Music Streaming Services and New Forms of Class Distinction." *Information, Communication & Society* 23, no. 13 (2020): 1909–1924.

Weisbard, Eric. *Top 40 Democracy: The Rival Mainstreams of American Music*. Chicago: University of Chicago Press, 2014.

Williams, Raymond. *The Long Revolution*. London: Chatto & Windus, 1961.

———. *Television: Technology and Cultural Form*. New York: Routledge, 1990. First published 1974.

Wodke, Larissa. "The Irony and the Ecstasy: The Queer Aging of Pet Shop Boys and LCD Soundsystem in Electronic Dance Music." *Dancecult: Journal of Electronic Dance Music Culture* 11, no. 1 (2019): 30–52.

Yochim, Emily Chivers, and Megan Biddinger. "'It Kind of Gives You that Vintage Feel': Vinyl Records and the Trope of Death." *Media, Culture & Society* 30, no. 2 (2008): 183–195.

Youngquist, Paul. *A Pure Solar World: Sun Ra and the Birth of Afrofuturism*. Austin: University of Texas Press, 2016.

Index

Abrams, Lee, 100–101, 125
Acland, Charles R., 18–19, 24, 75
Adelstein, Jonathan, 136
Adorno, Theodor, 16, 18, 58, 107
AdsWizz. *See* Pandora
AFTRA. *See* American Federation of Television and Radio Artists
Aglukark, Susan, 195
AirTran Airways, 88–89
album-oriented rock (AOR), 103–4
Alcatel, 21
Allocation for Music Producers Act (AMP Act), 149
Alpine, 21
Alt Nation, 221
Amazon, 151
Amazon Prime, 68
America Online (AOL), 91–93
American Blockbuster (Acland), 24
American Communications Act, 12
American Federation of Musicians, 145
American Federation of Television and Radio Artists (AFTRA), 144
American Honda, 21, 136
American Mobile Radio Corp, 31
American Mobile Satellite Corporation, 31
AMP Act. *See* Allocation for Music Producers Act

Anderson, Bill, 159
Android Auto, 210
AOL Music Live, 90–92
AOL Music Sessions, 91–92
AOL Radio, 150
AOL. *See* America Online
AOR. *See* album-oriented rock
Apollo Management, 30
Apple, 12, 151, 213–14: CarPlay, 210; iPod, 82, 91, 121, 196, 197, 204; Music, 67–68, 94–95, 131, 184, 214, 216–17. *See also* automobiles
Apples in Stereo, The, 172
Aquarium Drunkard (program), 172–74
Arbitron, 91
archetypes, hosts. *See* collector, archetype; tastemaker, archetype; veteran, archetype
Arditi, David, 67
Artist Confidential (program), 123, 150
artists, programming reach and, 108–11
Atlanta Braves, 23, 211
"Atlantic City" (song), 124, 129
Atlantic, The, 13, 171
Attali, Jacques, 69
audio ambience channels, 119–23
audio animators, term, 120
AudioID, 225
Audiovox, 21

293

294 INDEX

authenticity. *See* hybrid authenticity
automobiles, 61; privatizing automobility, 82–84; pursuing satellite radio subscribers in space of, 77–82; satellite radio infrastructure and, 36–38; SiriusXM's 360L radios in, 226; uncertain space of, 210
Azoff, Irving, 153

B. B. King, 161
B. B. King's Bluesville (program), 161
Backstreet Boys, 4, 105
Baker, Sarah, 139
Barnett, Kyle, 117
Baudrillard, Jean, 84–85
Beastie Boys, 111
Beatles, 1–2, 105, 155
Beats Electronics, 214
Beats Music, 150, 153
BeBe Winans, 159
Berland, Jody, 40, 58–59, 94, 190
Bertelsmann Music Group (BMG), 71
Best Buy, 21
Bezos, Jeff, 209
Biden, Joe, 170–71
Bijsterveld, Karin, 86
Billboard, 2, 4, 9, 28, 156, 219; article on German copyright law, 130; blurbs about music business news, 131; charts, 184, 200, 233n1; on country radio, 115; describing MySXM, 203; hearing outer space in music, 48–49; Kid Kelly, 215; on power and prominence of DJ, 155–56; profiling "future in record merchandising," 53; record label promotion reps polled by, 137; reflecting on Liberty Media, 211–12; reporting on mergers, 73; reporting on Pandora acquisition, 217–18; reporting on satellite transmissions, 41; reporting on Sirius and XM, 134, 147–48; review for "Stereo-Only Albums" in, 47; and satellite music radio, 22–25
Bitove, John, Jr., 192, 197–98
Blaakilde, Anne Leonora, 227
Black & Blue (album), 4–5
Blatter, Steve, 223
Blaupunkt, 21
Blink-182, 9
Blog Radio, 171–75
blogs, DIY aspect of, 171–75
BMG. *See* Bertelsmann Music Group
Bohemian Rhapsody (film), 111

Bolter, Jay David, 168
Bonini, Tiziano, 205
Bontec, 21
Bottomley, Andrew, 157, 213
Bourdieu, Pierre, 87
Bowie, David, 49, 51
Branson, Richard, 209
Bridgers, Phoebe, 174–75
Briskman, Robert, 30
Broadcasting, 46
Broadcasting Act, 189, 196
Broadcasting Magazine, 61
Broken Social Scene, 9, 192
Brooklyn Vegan, 172
Brooks, David, 171
Brooks, Garth, 9
Buffet, Jimmy, 108
Buried Treasure (program), 108–9
Burkart, Patrick, 213
Bürkner, Hans-Joachim, 106
burn, term, 139

cable, radio and, 75–77. *See also* television-radio parallels
Canada-U.S. Free Trade Agreement, 190
Canada, developing satellite radio in, 187; content circulation, 192–201; content regulations, 191–92; cultural lifelines, 201–7; North American Free Trade Agreement (NAFTA), 190–91; Sirius and XM moving into Canada, 188–92. *See also* content (Canadian), circulating
Canadian Broadcasting Corporation (CBC), 189–90
Canadian Independent Record Production Association (CIRPA), 194–95
Canadian Radio-television and Telecommunications Commission (CRTC), 188–89, 194–97
Canadian Satellite Radio (CSR), 191–92
Capobianco, Joe, 62
Carles.buzz, 171–72
cars, satellite radio in. *See* automobiles; mobility, privatization of
Casey Kasem's American Top 40, 120
Cashbox, 2
CBC Radio 3, 98, 192–94, 199–200. *See also* content (Canadian), circulating
CBC. *See* Canadian Broadcasting Corporation
CBS, 109–10
CD Radio, 34, 62–63
CDnow, 73

celebrities: collector archetype, 160–71; overview, 155–56; perceived value and, 107–8; piloted listening, 181–83; programming reach and, 108–11; radio and sociability, 157–58; stars on air, 158–60; tastemaker as archetype, 171–75; veteran as archetype, 175–80. *See also* nostalgia, drawing on
Chaitovitz, Ann, 144
channels: artists and celebrities, 108–11; asserting value by commercial means, 109–10; Beatles Channel, 2, 113; Classic Vinyl, channel, 120, 122; communicating perceived value, 107–8; country music format, 115–16; decades music channels, 119; distinction and expansiveness in channel lineup, 100–106; duality of value, 110–11; genre category breakdown, 95–98; Grateful Dead Channel, 113; initial program lineups, 98–100; life cycles of music, 124–28; music having value beyond service, 1067; No Shoes Radio, channel, 222; perceived value on channel lineup, 106–12; and piloted listening, 181–83; pop-up channels, 110–11, 222; The Pulse, channel, 101; Radio Margaritaville, channel, 108; Shade 45, channel, 109; streaming metaphor, 95–96; and subgenres, 112–13; superserving formats, 114–15; term, 96
Channels You Might Like (feature), 230
Chapman, Tracy, 90
Chevry, Bernard, 53
Child's Introduction to Outer Space, A (album), 48
Chill Trending Tracks-Powered by YouTube (playlist), 223
Chopra, Naveen, 218
Chow, Rey, 66–67
Chronicles (Dylan), 164, 166, 168–69
Circuit City, 21
circulation, 14–17
circulatory paths, 14–15
circulatory turn, 14–15
CIRPA. *See* Canadian Independent Record Production Association
cities, mobility in, 89. *See also* automobiles
Clarion, 21
Clark, Brandy, 215
Clarke, Arthur C., 29
CLASSICS Act. *See* Compensating Legacy Artists for Their Songs Service and Important Contributions to Society Act
Clear Channel Communications, 21, 31, 62

Clear Channel, 12, 103–4
Clinton, George, 49, 108
Close to The Edge (show), 111
clustering, 120–21
CMJ New Music Report, 40–41, 172
Coddington, Amy, 114–15
coffee shops, potential subscribers in, 89–90
Colbjørnsen, Terje, 96
Coldplay, 123
collector, archetype, 156–57 (*see also* celebrities); collector-as-expert, 160–62; freeform radio format, 166–67; hybrid authenticity, 163–65; and old time radio (OTR), 167–71; radio voice, 164–66
Coltrane, Alice, 174
Columbia Capital, 21
Columbia House, 70–74
Columbia Record Club, 70–74
commodity scientism, 48
Compensating Legacy Artists for Their Songs Service and Important Contributions to Society Act (CLASSICS Act), 149, 151
Conference on Space Communication, 38
consolidation, 4, 12, 18, 26, 62, 132, 136
consumer technologies, 47–48
content (Canadian), circulating: benefits of licensing satellite radio, 195; CBC channel removal, 199–201; licensing process, 194–95; mobile listening devices, 196–97; outlining licensing framework, 196; SiriusXM Canada becoming more American, 197–99
Conyers, John, 134
Cookman, Tomas, 203
Copps, Michael, 136
Copyright Royalty Board (CRB), 141–42, 149
country music format, investing in, 115–16
Country Music Hall of Fame Hour, The, 123–24
Country Music Hall of Fame, partnership with, 123–24
Country Radio Seminar, 64, 115
Cowie, Zach, 173–74
Cracked Rear View (album), 71
Crave, 68
CRB. *See* Copyright Royalty Board
Creed, 105
Crooked Media, 225
CRTC. *See* Canadian Radio-television and Telecommunications Commission (CRTC)
Cruisin' America (program), 176
Crupnick, Russ, 229

CSR. *See* Canadian Satellite Radio
Cultish, 68
cultural lifelines, Canadian satellite radio and, 187, 201–7
cultural technology, satellite as, 40–44
Cumulus Radio, 216
curation, term, 158
"Curves of the Needle, The" (Adorno), 107

dailiness, 227
Dark Side of the Moon (album), 50
Dave Matthews Band, 9
"Dead Disco" (song), 126
Dear Johnny (album), 174
decades music channels, 119
Deleuze, Gilles, 121
Delphi-Delco Electronics, 21
Department of Justice, USA, 135–36
Destiny's Child, 9
devalue, music, 139–42
devaluing, music, 130–31
Diamond the Artist, 178
Diamond, Israel, 69
Digital Media Association (DiMA), 150–51
Digital Millennium Copyright Act (DMCA), 130, 143
Digital Music Co., 70
Digital Performance Right in Sound Recordings Act (DPRA), 144
Dijck, José van, 205
DiMA. *See* Digital Media Association
DirecTV, 21, 31, 74, 138
Disney+, 68
distinction, satellite radio channel lineup, 100–106
Dixon, Neil, 194
DIY. *See* do-it-yourself
DMCA. *See* Digital Millennium Copyright Act
do-it-yourself (DIY), 172–73
Dolan, Peter, 51
Donnelly, Patrick, 217
Doucet, Luke, 192
Douglas, Susan J., 51
Dr. Dre, 105
Drott, Eric, 24
Drowned in Sound (blog), 126
DRPA. *See* Digital Performance Right in Sound Recordings Act
Duffy, Brooke Erin, 205
Dupri, Jermaine, 108
Dylan, Bob, 161; freeform radio format, 166–67; and old time radio (OTR), 167–71;

radio voice, 164–66. *See also* collector, archetype

Earthing Project, 44
EchoStar, 138
EDM. *See* electronic dance music
electronic dance music (EDM), 9
electronic music, investment in, 116–17
Eliot, T. S., 167
Ellington, Duke, 156
Ellis, Jason, 220
ELV1S: 30 #1 Hits (album), 1–2
Elvis Radio, 2
embedded radio. *See* embeddedness, expansion and
embeddedness, expansion and, 226–28
emerging, term, 200
Eminem, 4, 105, 108
Eminem Show, The (album), 159
Empires of Entertainment (Holt), 135
Encoda System, 36
Epilogues, The, 172
expansion, SiriusXM: content circulation, 192–201; overview, 184–87; Sirius and XM moving into Canada, 188–92. *See also* limits, expansion
expansiveness, satellite radio channel lineup, 100–106

FCC. *See* Federal Communications Commission
Federal Communications Commission (FCC), 12, 108, 136, 188; and licensing satellite radio, 38–40
Feel What U Feel (album), 180
Feingold, Russ, 132
Feist, 192
Fisher, Mark, 12, 125
fluidity, 19–20, 87–88, 95–96, 132
FM-1, satellite, 33–34
FM-2, satellite, 33–34
FM-3, satellite, 33–34
FM-4, satellite, 34
FM-5, satellite, 34
FM-6, satellite, 34
Folkways Records, 48
formats, superserving, 114–15
Formula One, 23, 211
Fortune, 35
Fox News, 220
Fraunhofer Institute, 21
Frear, David J., 141
free music, 139–42

freemium, 140, 230
Frith, Simon, 17, 106-7, 208
From His Home to Yours (program), 170
Fujitsu Ten, 21
Future of Music (report), 12

Gandini, Alessandro, 205
Gates, Kevin, 48
Geechie Dan, 178
General Motors, 21, 36, 78, 136
generational belonging, 127
Genmar Holdings Inc., 89
genre categories, breaking down streams into, 95-99
GEO. *See* geosynchronous equatorial orbit
geostationary, term, 33
geosynchronous equatorial orbit (GEO), 33, 55, 57-58
geosynchronous, term, 33
Gervais, Ricky, 220
"Gettaway" (song), 10
Gill, Alexandra, 193
Gitelman, Lisa, 19
Glazier, Mitch, 150-51
global assault, phrase, 4-5
global hit music, 3-5
Global Radio, 3
Globalization (album), 185
Globalization Radio, 185-87
Globe and Mail, 46, 193
Godrich, Nigel, 40
Goodlatte, Bob, 149
Goodman, Fred, 104-5
Google, 151
Google Chromecast, 210
Gorilla vs. Bear Blog Radio, 172, 174
Graham, Stephen, 89
Grandmaster Flash, 159
Grateful Dead, 2
Green, Irving B., 53
Greenstein, Scott, 17, 160, 215, 218-20, 222
Greybull, Wyoming, Sirius ad for, 53-54
Griffin, Meg, 177
Grohl, Dave, 210
Grusin, Richard, 168
Gulf Wars, 66

Haines, Emily, 126
Hair Nation, 113
handcrafted programming, 160
"Happy Together" (song), 147
Harris, Emmylou, 159
Harrison, Chris, 150-51

Hart, Kevin, 210, 220
Hartman, Sara, 48
Harvey, Eric, 95
Healey, Jeff, 195
"Heartbreak Hotel" (song), 2
Heidegger, Martin, 66
Henley, Don, 132, 159
Hesmondhalgh, David, 107
Hilmes, Michele, 2, 185
Hipster Runoff Blog Radio (program), 120
Hollywood Reporter-Billboard Media Group, 25
Hollywood Reporter, The, 138
Holt, Jennifer, 18, 135
Home Music Store, 70
Hootie & the Blowfish, 71
Horkheimer, Max, 18
hosts, celebrities as, 158-60. *See also* celebrities
hotels, potential subscribers in, 90-91
"How to Construct a Time Machine" (essay), 50
Howard, Phillip N., 227
Huber, Alison, 139
Hughes Electronics, 36
Hughes Network Systems, 21
Hughes Space and Communications, 21
Hulu, 68
Hunt, Lynn, 11
Hunt, Sam, 215
Hutcheon, Linda, 14
hybrid authenticity, 163-65
hybridity, 19-20, 164
hypermobility, 206-7
hypertargeting, 222, 225
Hyundai Autonet, 21

"I Wish I Was the Moon" (song), 10
I'm Not There (film), 164, 165
identity, music and. *See* embeddedness, expansion and
iHeartMedia, 216, 220
In Search of Space (album), 50
In the Aeroplane Over the Sea (album), 10
infrastructures, 42, 77, 137, 190, 223
Innis, Harold, 58-59
instantaneous communication, 206-7
"Intergalactic" (song), 10
International Bluegrass Music Association
International Telecommunications Union (ITU), 38, 55, 190
Iovine, Jimmy, 109
iPod, , 82, 91, 121, 196, 197, 204. *See also* Apple

ITU. *See* International Telecommunications Union
iTunes Store, 12, 215. *See also* Apple

Jackson, Janet, 9
Jackson, Michael, 111
Jameson, Fredric, 14, 122
Jazz Alive (program), 22
Jenkins, Henry, 19
Jepsen, Carly Rae, 88
Jericho (album), 129
"Jesus, Etc." (song), 230
Jewel, 90
Jewel, Johnny, 174
Jewell, Katherine Rye, 102
Johnson, Aaron J., 156
Jones, Mike, 139
Jones, Quincy, 45, 108, 159
Judiciary Hearing on Protecting and Promoting Music Creation for the 21st Century, 149
Juno Awards, 72

Kaplan, Caren, 66
Karmazin, Mel, 81, 134–38, 146–47, 184, 207, 211
kbps. *See* kilobits per second
Kelly, Megyn, 220
KEXP (Seattle), 64
Kid Kelly, 215
kilobits per second (kbps), 41
Kirkpatrick, Bill, 21, 77
Kornbluh, Anna, 15
Kraft Music Hall (program), 180
Kramer, Michael J., 102–3
Kulawick, Geoff, 194

labels, direct licenses with, 146–48
Lacey, Kate, 157
Ladies and Gentlemen We Are Floating in Space (album), 10
Lambert, Josiah, 172
Lange, Bastian, 106
Lanois, Daniel, 166
LaPolt, Dina, 151
Las Vegas Consumer Electronics Show, 101–2
Last.fm, 150
LCC International, 21
Lee, Abrams, 120–21, 159
Lehman Brothers, 30
Leo G, programmer, 115
LEO. *See* low earth orbit

Levy, Joe, 162
Liberty Media, 6, 12, 138, 152, 184, 229; as conglomerate; 211–13; and story of satellite radio, 23–24. *See also* Sirius; SiriusXM; XM
licensing, satellite radio, 38–40
life cycles, music, 124–28
Lil Wayne, 50–51
limits, expansion: Liberty Media conglomerate, 211–13; overview, 208–11; Pandora acquisition, 216–21; pursuing new avenues, 224–31; radio industry-record industry connection, 221–24; subscription music radio in streaming space age, 213–16. *See also* Canada, developing satellite radio in; expansion, SiriusXM; Liberty Media; Sirius; SiriusXM; XM
Lindy, Scott, 115–16, 124
listeners, targeting, 13–14, 17–18
Live from the Underground (Jewell), 102
Live Nation, 12
Live Nation Entertainment, 211–13
LL Cool J, 161, 177–78, 180, 210
Loeb, Lisa, 178–79
Looking Backward (Bellamy), 95
Lopez, Jennifer, 9
Los Angeles Times, 70, 162
Love and Theft (album), 162, 164
"Love Me Do" (song), 2
low earth orbit (LEO), 33
Lucent, 21
Lululemon, 68
Lynch, Roger, 17, 218

Maffei, Greg, 151
Magaudda, Paolo, 205–6
Malone, John, 138, 211
Manuel, Richard, 129
Margolese, David, 30, 78
Marketing, 199
Marklund, Petra, 48
Marley, Bob, 2
Marsalis, Wynton, 159
Marshall Mathers LP, The (album), 4
Marshall, Lee, 163
Martin, Kevin, 136
Marvel, 220
Maxar Technologies, 209–10
MCA Music Entertainment Canada, 72
McCartney, Paul, 75, 123
McCoury, Del, 108
McDowell, Robert, 136

McGraw-Hill Building, 34–35
McLean, Bethany, 35, 45–46
McLeod, Ken, 49, 117–19
Media Industries (Journal), 18
Meier, Leslie M., 141
Meister, William von, 70
Memphis Minnie, 165–66
Mercury Record Corporation, 53
Metric, 125, 192
Meyer, James E. (Jim), 152, 207, 211
Meyers, Cynthia B., 180
micro-genres, 113–12
Midrange Vancouver (blog), 12
Millennium (album), 4
Mitsubishi, 21
MMA. *See* Music Modernization Act
mobile listening, 201–7
Mobile-One, 21
mobility, privatization of: adapting and appeasing online listeners, 93–94; AOL-XM partnership, 91–93; beyond vehicle, 88–89; characterizing ideal subscriber, 82–83; class and cultural hierarchies, 83–86; coffee shops, 89–90; critiquing mobility as new universalizing condition, 87–88; cultural binaries of high and low culture, 88; explanations, 86; extensive systems of immobility, 86–87; hotels, 90–91; new mobilities paradigm, 86–87; Sirius-Sprint partnership, 93; subscribers in urban areas, 89; targeted places and spaces for ideal subscribers, 91
Modern Times (album), 162, 164, 168
MOJO, 50
Monk, Charlie, 124
Montell, Amanda, 68
Moon & Antarctica, The (album), 10
Moon Safari (album), 10
Mooradian, Mark, 69
Morris, Jeremy, 95, 205
Morrow, "Cousin Brucie," 175–76
Motorola, 21
Mr. Worldwide. *See* Pitbull
Mraz, Jason, 90
Mull, Amanda, 13
Mullen, Megan, 76–77
Murdock, Graham, 132
Murphy, Paul E., 46
Music Choice, 150
Music Choice Europe, 70
Music for Heavenly Bodies (album), 52
Music from Outer Space (album), 47

Music Modernization Act (MMA), 130; discussion preceding Senate approval of, 149–52; origin of, 148–49
music programming. *See* programming, perceived value and satellite streams in
MusicMatch, 68
MusicNet, 68, 69
MusicWatch, 229
Musk, Elon, 209
My Old Kentucky Blog, 171
MySXM, 203, 207

NAB. *See* National Association of Broadcasters
Nacional Records, 203
NAFTA. *See* North American Free Trade Agreement
Napster, 4, 91–92
narrowcasting, 76
NASCAR, 81–82
Naser, Rida, 182
Nashville, Tennessee, foothold in, 123–24
National Association of Broadcasters (NAB), 39
National Association of Music Merchants, 46
National Post, 133
Nebraska (album), 124
negative option billing, 71
Negus, Keith, 109, 162
Nelly, 105
Nelson, Willie, 124
Netflix, 68
NetRadio, 150
new media, 9, 19–20, 45, 62, 132, 168, 188, 221, 227
New Musical Express, 2
New York Journal-American, 175
New York Magazine, 58
New York Times, The, 170, 180, 206, 220
Newport Folk Festival, 163
Nieborg, David, 205
Nine Inch Nails, 48
No Direction Home (film), 164
No Strings Attached (album), 4
Noise: The Political Economy of Music (Attali), 69
non-music programming, 141–42
North American Free Trade Agreement (NAFTA), 190–91
nostalgia, drawing on: audio ambience channels, 119–23; decades music channels, 119; defining commercial radio stations, 113;

nostalgia (*continued*)
electronic music format, 116–17; format investment examples, 115–17; justifying cost subscription access, 117–19; life cycles, 124–28; micro-genres, 113–12; Nashville footholds, 123–24; organization of self, 112; superserving formats, 114–15
NSYNC, 4, 105

O'Brien, Conan, 220
Obelysk Media, 199
OEM. *See* original equipment manufacturer
Oh Mercy (album), 166
old time radio (OTR), 167–71
1 (album), 1–2
one, number, 1
Oobu Joobu (radio show), 75–76
Oops!... I Did It Again (album), 4
orbit: central concept, 13–15; conceptualizing music culture, 56–59
original equipment manufacturer (OEM), 78–79
OTR. *See* old time radio
Our World, 55–56
outer space, music radio and: hearing outer space, 48–50; licensing satellite radio, 38–40; and limits of expansion, 224–31; music communicating space, 44–56; orbits, 56–59; overview, 28–29; radio into space, 29–38; satellite as cultural technology, 40–44. *See also* satellite, object; space, music communicating

PAC v4 Audio Codec, 41
Paglen, Trevor, 42
Panasonic, 21
Pandora, 12, 146, 150, 151; SiriusXM acquiring, 216–21. *See also* limits, expansion; outer space
Panero, Hugh, 81, 91
Pareles, Jon, 164
Parker, Mimi, 49
Parks and Recreation, 10
Parks, Lisa, 41, 66
Parquet Courts, 215
Parsons, Gary M., 136
partnerships, industry, 21–22
Paul, Brad, 137
Peacock, 68
Pearl Jam, 2, 155
Peer-Southern, 69
perceived value, 13–14; on channel lineup, 106–12; of music, 15–16, 18–19

performance rights organizations (PROs), 142–43
Perren, Alisa, 18
Perry, Lee, 49
personalities, radio: collector, 160–71; tastemaker, 171–75; veteran, 175–80. *See also* celebrity radio voice
personality music radio, 155. *See also* celebrity radio voice
Peters, John Durham, 42, 157
Peters, Mike, 114
Petty, Tom, 108–9, 161
Phelps, Phlash, 7
Philco transistor radios, 47–48
Phish, 2
Piggies, The, 172
piloted listening, 181–83
Pink Floyd, 50
Pioneer, 21
Pitbull, 185–87
Pitchfork (blog), 10, 126–27
Platform Society, The (Dijck), 205
platformization, 205, 211–14
platforms, integrating, 222–23
Poell, Thomas, 205
political economy, context of: music as rationalization, 139–42; Music Modernization Act, 148–54; overview, 129–33; royalty problem, 142–48; Sirius-XM merger, 133–38
pop-up channels, 19, 105, 110–11, 222
postmodernism, periodization of, 122
Powers, Devon, 95
premium radio/listeners: altering accessibility of radio, 64–66; consolidation and capitalization, 61–62; emphasis on change, 62–63; satellite technology and, 66–67; themes of exceptionality, newness, and distinction, 64
Presley, Elvis, 1–2
Pressplay, 68
Prey, Robert, 216, 222–23
Priestman, Chris, 157–58
Prime 66 Partners, 30
programming, perceived value and satellite streams in: channel breakdown, 95–100; channel lineup perceived value, 106–12; distinction and expansiveness, 100–106; drawing on nostalgia, 112–24; life cycles of music, 124–28; recognizing enormous power of reach, 110
Project Moon (album), 51
PROs. *See* performance rights organizations

Public Knowledge, 135
Punk Rock Blitzkrieg: My Life as a Ramone (Ramone), 180

Queen, 110–11
Queen Latifah, 159

racialized others, 45
radio: defining "premium" radio, 61–67; licensing, 38–40; lineup expansiveness, 100–106; national to transnational, 188–92; nostalgia and music heritage on, 112–24; pursuing subscribers, 77–82; and record industry, 221–24; satellite music radio, 22–27; SiriusXM acquiring Pandora, 216–21; sociability and, 157–58; into space, 29–38; from space age to satellite radio, 44–56; as stealth medium, 185; in streaming space age, 213–16; television on, 74–77. *See also* satellite radio
Radio.com, 150
Radio (album), 177–78
Radio & Records, 25
Radio 3. *See* CBC Radio 3
radio into space: automobiles and, 36–38; companies establishing footprint in USA, 34–36; companies launching satellites, 30–33; satellite as object, 29–30; XM satellite, 33–34
Radio Shack, 21
Radiosat 1–3, 33
RAIN News, 25, 110, 210
Raitt, Bonnie, 108, 123, 159
Ramone, Marky, 179–80
rationalization, music as, 139–42
Razlogova, Elena, 173
RCA, 109–10
RCA Records, 1
Rebich, Larry, 74
record clubs, 70–74
record industry, radio relationship with, 221–24
Record Retailer, 2
Record, Chris, 225–26
Recording Industry Association of America (RIAA), 3–4, 144–45
records, orbits and, 57
Recovering the Satellites (album), 102
Redmond, Mark, 198
Reed, Lou, 49
Regan, Jeff, 221
Regev, Motti, 106
Rehr, David, 134–35

remediation, defining, 168
remote sensing, term, 66
Reputation (album), 222
Return to Saturn (album), 10
Rexha, Bebe, 210
Reynolds, Simon, 11, 119, 158
RIAA. *See* Recording Industry Association of America
Richardson, Mark, 10
Richman, Jonathan, 17
Rimes, LeAnn, 159
Rise Against, 48
Rise and Fall of Ziggy Stardust and the Spiders from Mars, The (album), 51
Robinson, Smokey, 151
Rock and Roll Hall of Fame Museum, 176–77
"Rock around the Clock" (song), 56
Rock the Bells Radio, 178
Roessner, Jeffrey, 113, 158
Rogan, Joe, 220
Rogers, Ian, 153
Rolling Stone magazine, 102–4, 162, 168, 231
Rosen, Hilary, 132
Rothblatt, Martine, 30
Rounder Records, 137
Royal Commission on Radio Broadcasting, 189–90
royalties, problem of: direct licenses with labels, 146–48; performance rights, 142–43; rights organizations, 143–46
royalty rates, determining, 129–33
Ruff, Willie, 46–47
Rushkoff, Douglas, 7–8
Russo, Alexander, 21, 77
Ryan, Mary Pat, 83

Saddest Factory Radio (program), 174
Samara, Noah, 3
San Francisco Opera House, 41
Santana, 105
Santiago, Felipe Perez, 44
Sanyo, 21
Satellite CD Radio Inc., 30
satellite digital audio radio service (SDARS), 3, 39
satellite music radio, 22–27
"Satellite of Love" (song), 49
satellite radio: celebrity radio voice on, 155–83; central concepts regarding, 13–14; and circulatory paths in music, 14–15; claims about superiority of, 9; combining mecha approach to, 20; connections between outer space and, 28–59; distinction and

satellite radio (*continued*)
expansiveness in channel lineup, 100–106; earliest musician-hosts of, 116; expansion of, 184–207; first attempt establishing, 3; and hybridity, 20; and industry partnerships, 21–22; as integral component of listening contexts, 17–18; legacy broadcast media and, 12–13; licensing, 38–40; and limits of expansion, 208–31; and mobile listening, 201–7; music culture during launch of, 3–5; perceived value and satellite stream in music programming, 95–128; political economy shaping, 129–54; as product of times, 11–12; pursuing subscribers of, 77–82; radio industry–record industry connection, 221–24; royalty problem of, 142–43; and satellite music radio, 22–27; study of, 17–18; targeting subscribers, 60–94; and technology of satellite, 19–20. *See also* Sirius; SiriusXM; XM

satellite, object, 29–30; as cultural technology, 40–44; grounding, 43–44; as mythic force, 46–47

Savage, Jon, 50

Scannell, Paddy, 14, 157, 227

Scott, A. O., 165

Screen Actors Guild–American Federation of Television and Radio Artists, 145

SDARS. *See* satellite digital audio radio service

Search for Extraterrestrial Intelligence (SETI), 44

Sears, 21

Selling Sounds (Suisman), 208

Serial Productions, 220

SETI. *See* Search for Extraterrestrial Intelligence

Shaggy, 9

"Shape of My Heart" (song), 4

Sharp, 21

Sheller, Mimi, 87

Sheridan Broadcasting Network, 76

Shumway, David R., 125

signal traffic, 41–42

Silverman, Tom, 217

simulacrum, concept, 121

Sing Out!, 137

SIRIUS-Seek, 93

Sirius, 111; 100 streams of satellite radio, 97; audio ambience channels, 119–23; being distinct from commercial broadcast radio, 9–10; celebrity hosts for, 158–69; comparisons, 8; country music format, 115–16; decades music channels, 119; early advertising campaigns for, 6–7; early life of, 29–59; electronic music format, 116–17; foothold in Nashville, 124; harnessing power of orbits, 56–59; and industry partnerships, 21–22; and licensing satellite radio, 38–40; and life cycles, 124–28; merging with XM, 133–38; moving into Canada, 188–92; music communicating space, 44–56; NASCAR partnership, 81–82; notion of progress promoted by, 122–23; offering multiple versions of one format, 104; partnering with Sprint, 93; perceived value and satellite stream in music programming, 95–128; perceived value on channel lineup, 106–12; pursuing new avenues, 224–31; pursuing satellite radio subscribers, 77–82; recognizing enormous power of reach, 110; Rock and Roll Hall of Fame Museum partnership, 176; royalty problem, 142–48; satellite as cultural technology, 40–44; and satellite music radio, 22–27; targeting subscribers, 60–94; and television industry, 74–77; and value of music, 139–42. *See also* SiriusXM; XM

SIRIUS Canada (Sirius Canada), 191–92. *See also* Canada, developing satellite radio in

Sirius Satellite Radio, 2

Sirius XM Canada, 186, 196–200. *See also* Canada, developing satellite radio in

Sirius XM Canada Holdings Inc., 198–99. *See also* Canada, developing satellite radio in

Sirius-1, satellite, 31–33

SiriusXM, 2–3, 6, 68, 111, 125, 131; changes in leadership at, 211; channels of, 98–100; competing with Apple Music, 214; debuting SiriusXM 3.0, 202–7; and embeddedness, 226–28; expansion of, 184–207; generating synergy of, 222–23; and Globalization Radio, 185–87; and limits of expansion, 208–31; mass and niche listener preferences, 17–18; and Maxar Technologies, 209–10; merged channel lineup, 105–6; merger history, 133–38; and mobile listening, 201–7; and Music Modernization Act, 148–54; new LA studio of, 219–20; opposing CLASSICS Act, 151–52; Pandora acquiring, 216–21; as part of Liberty Media conglomerate, 211–13;

INDEX

partnering with platforms, 221–24; piloted listening, 181–83; pop-up channels, 110–11; royalty problem, 142–48; royalty problem of, 142–43; satellite as cultural technology, 40–44; and satellite music radio, 22–27; and subscription music radio in streaming space age, 213–16; and value of music, 139–42. *See also* Sirius; XM
SiriusXM 3.0, 202–7
SiriusXM Holdings Inc., 212
SiriusXMU, 113, 120, 123, 171–75
Skeggs, Beverley, 87
sky media, 42–43
Slaight Communications, 199
Slate, 173
Slim Cessna's Auto Club, 172
Smith-Shomade, Beretta E., 76–77
Smith, Darren, 222
Smith, Elliott, 48
Smith, Lamar, 135
Smith, Patti, 100
Snoop Dogg, 108
sociability, 157–58
Sohn, Gigi, 135
Soho Weekly News, The, 175
Songza, 150
Sony, 21
Sony Music Entertainment, 72–73, 151
Sony Music Group, 213
SoulCycle, 68
Sound and Safe (Bijsterveld), 86
Sound Recording Act, 142, 144
sound souvenirs, 121
SoundCloud, 220, 222–23
SoundExchange, 142, 143–46, 152
sounds, recycling, 127. *See also* nostalgia, drawing on
space capitalism, 208–9
"Space Lord" (song), 10
space rock, advent of, 49–52
space, music communicating: advent of space rock, 49–52; borders and boundaries, 55; crystallizing ephemeral sound, 44–45; hearing outer space, 48–50; idealistic aspirations, 52–53; imagining ideal consumer, 52; impossible distances, 44; and new consumer technologies, 47–48; praise for satellite radio services, 45–46; revolutionary ideas about collapsing time, 55–56; satellite as mythic force, 46–47. *See also* outer space, music radio and; satellite, object

SpaceX, 209
Sparhawk, Alan, 49
Spears, Britney, 4, 105
Spilker, Hendrik Storstein, 96
Spinning Indie, 172
Spoon, 48
Spotify, 12, 21, 68, 94, 131, 140, 146, 151, 169, 184, 205, 214, 216, 217
Springsteen on Broadway, 170
Springsteen, Bruce, 124, 129, 161, 170
Sprint, 93
Sputnik, launch of, 42–43, 49
Squeezebox Duet, 204
Standley, Johnny, 121
Star, Susan Leigh, 226–27
Starbucks, 68, 89–90
Starosielski, Nicole, 41
Stay with Lisa Loeb (program), 178–79
Stern, Howard, 7, 79–81, 111, 138, 156, 219
Sterne, Jonathan, 87
Stiletto 100, 93
Stiletto 2, 204
Sting, 159
Stingray Digital, 70
Stitcher, 220
STMicroelectronics, 21
Straw, Will, 15
streaming: "free" music and, 139–42; metaphor, 95–100; music subscription before, 67–74; space age of, 12–22; subscription music radio in age of, 213–16. *See also* subscription models, shift to
Streaming Music, Streaming Capital (Drott), 24
streamlining, term, 132–38
Street, Dusty, 176–77
Streeter, Thomas, 94
structural convergence, term, 135–36
Student Exchange Program, 172
subgenres, 113–14
subscriber-listeners, 26, 61, 65, 67, 87, 112, 202. *See also* subscribers, targeting
subscribers, targeting: automobility/mobile privatization, 82–94; growing subscribers, 22, 185; orbit metaphor, 68; overview, 60–61; potential subscribers, 89–90, 185, 188, 203, 221, 223 premium listers for premium radio, 61–67; pursuing satellite radio subscribers, 77–82; record clubs, 70–74; shift to subscription models for music consumption, 67–74; subscription efforts, 68–70; television on radio, 74–77.

subscribers (continued)
    See also automobiles; coffee shops, potential subscribers in; hotels, potential subscribers in; mobile, privatization of
subscription models, shift to, 67
subscription satellite radio, 22, 163, 183, 208
subscriptions: before streaming, 67–74; in streaming space age, 213–16. See also subscribers, targeting
Swedish Egil, DJ, 117
Sweeney, Terrence, 31–33
Swift, Taylor, 222
SXM Media, 225
SXM Podcast Universe, 225
SXM-7, satellite, 34

Tacchi, Jo, 127
talk programming, shift toward, 225
Tarter, Jill, 44
tastemaker, archetype, 156–57. See also celebrity radio voice, 171–75
Tate, Deborah Taylor, 136
Taylor, Timothy D., 107
Taylor, Tony, 44
TCI. See Tele-Communications, Inc.
Team Coco, 220
Team Coco Radio, 220
technology, evaluation of, 228–29
Telcom Ventures, 21
Tele-Communications, Inc. (TCI), 23
Telecommunications Act, 12, 21
Telesat, 21
Television Digest with Electronics Reports, 45–46
Television Magazine, 46
television-radio parallels, 74–77
"Telstar" (song), 48–49
Tepper, Jeremy, 116
Terrence, Gregg, 195
Theme Time Radio Archive, 161
Theme Time Radio Hour (TTRH), 76, 161–62; "Baseball" (episode), 168–69; freeform radio format, 166–67; "Friends & Neighbors" (episode), 166–67; and old time radio (OTR), 167–71; radio voice, 164–66
Thorburn, David, 19
Thundercat, 48
Timbaland, 10–11
Time Out of Mind (album), 164
time, collapsing, 55–56
Tool, 9
Toronto Star, 199

Tortured Poets Department, The (album), 222
Toshiba Records, 47
Toynbee, Jason, 139
Trump, Donald, 170–71
"Try Again" (song), 10–11
Tsing, Anna, 15
TTRH. See *Theme Time Radio Hour*
Tull, Jethro, 53
Turner, Tina, 2
Turquoise Wisdom, 173–74
Turtles, 147–48
TV industry, radio and, 74–77
Tweedy, Jeff, 230
twenty-first century, satellite radio and record industry in, 221–24
2Pac, 9

U2 X-RADIO, 111
Universal Music Group, 111, 151, 213
Universal Music Publishing, 111
Urry, John, 87

Valle, Marc Esteve Del, 216
value. See perceived value
Van Halen, 153
Van Zandt, Steven, 116
Variety, 25
Veal, Michael, 50
Vedder, Eddie, 155
vehicles, radio in. See automobiles; mobility, privatization of; subscribers, targeting
Verge, The (program), 200
veteran, archetype, 156–57. See also celebrity radio voice, 175–80
Viacom, 12
Victor Talking Machine, 109
Visteon, 21
vocal delivery, 164–66. See also collector, archetype; Dylan, Bob
Voices of the Satellites! (album), 48
Vulture, 10
Vuolo, Art, Jr., 6–7

Waal, Martijn de, 205
Wall Street Journal, 134
Warner Music Group, 72–73, 151–52, 213
Warner, Andrea, 194
Warner, Kristen, 77
Werde, Bill, 103
Weiner, Jonah, 173
Weisbard, Eric, 114
Westergren, Tim, 217

Wharton, Dennis, 134
What About Now (album), 222
"What Makes You Beautiful" (single), 215
Wheeler, Cleveland, 7
Williams, Raymond, 18, 65
Wilson, Brian, 159
Wilson, Carl, 10
Winfrey, Oprah, 138
Wireless World, 29
Witz, Jennifer C., 105, 211, 225
Wolff, Michael, 58
WorldSpace Satellite Radio, 3, 172
Wyatt, Robert, 174

xenon ion propulsion system (XIPS), 34
XIPS. *See* xenon ion propulsion system
XM, 111; audio ambience channels, 119–23; being distinct from commercial broadcast radio, 9–10; celebrity hosts for, 158–69; comparisons, 8; country music format, 115–16; decades music channels, 119; early advertising campaigns for, 6–7; early life of, 29–59; electronic music format, 116–17; foothold in Nashville, 124; harnessing power of orbits, 56–59; and industry partnerships, 21–22; initial program grid of, 98; launching satellites into space, 33–34; and licensing satellite radio, 38–40; and life cycles, 124–28; merging with Sirius, 133–38; moving into Canada, 188–92; music communicating space, 44–56; notion of progress promoted by, 122–23; offering multiple versions of one format, 104; partnership with Country Music Hall of Fame, 123–4; perceived value and satellite stream in music programming, 95–128; perceived value on channel lineup, 106–12; pursuing new avenues, 224–31; pursuing satellite radio subscribers, 77–82; recognizing enormous power of reach, 110; royalty problem, 142–48; satellite as cultural technology, 40–44; and satellite music radio, 22–27; satellites launched by, 33–34; targeting subscribers, 60–94; teaming up with AOL, 91–93; and television industry, 74–77; and value of music, 139–42. *See also* Sirius; SiriusXM
XM Canada, 55
XM Hear Music Series, 90
XM Satellite Radio, 3, 30–31, 35–36
XM Satellite Radio Holdings, 192
XMU Sessions, 123

Y2K problem, 4, 11
Yahoo! Music, 150
Yo Soy Juanes (program), 203
Yorke, Thom, 140
YouTube, 151, 153, 222–23;
YouTube 15 (program), 17

Zwerwer, Leslie, 216

Founded in 1893,
UNIVERSITY OF CALIFORNIA PRESS
publishes bold, progressive books and journals
on topics in the arts, humanities, social sciences,
and natural sciences—with a focus on social
justice issues—that inspire thought and action
among readers worldwide.

The UC PRESS FOUNDATION
raises funds to uphold the press's vital role
as an independent, nonprofit publisher, and
receives philanthropic support from a wide
range of individuals and institutions—and from
committed readers like you. To learn more, visit
ucpress.edu/supportus.